Inorganic Nitrate and Human Body Homeostasis

Songlin Wang
Editor

Inorganic Nitrate and Human Body Homeostasis

With Contributions by
Jian Zhou and Lizheng Qin

Springer

PMPH PEOPLE'S MEDICAL PUBLISHING HOUSE

Editor
Songlin Wang
Beijing Stomatological Hospital
Capital Medical University
Beijing, China

with contributed by
Jian Zhou
Beijing Stomatological Hospital
Capital Medical University
Beijing, China

Lizheng Qin
Beijing Stomatological Hospital
Capital Medical University
Beijing, China

ISBN 978-981-95-2823-3 ISBN 978-981-95-2824-0 (eBook)
https://doi.org/10.1007/978-981-95-2824-0

This Springer imprint is published by the registered company Springer Nature Singapore Pte Ltd.
The registered company address is: 152 Beach Road, #21-01/04 Gateway East, Singapore 189721, Singapore

Preface

Throughout human history, many things have been misunderstood, and nitrate is one of them. Cancer has long been a source of great fear, and for a long time, nitrates and nitrites were believed to have carcinogenic effects. In China, the public has developed a deeply rooted belief that nitrates and nitrites are harmful to human health. This perception has profoundly influenced people's views on health and lifestyle and even affected the broader social and economic context.

In fact, from ancient times to the present, human diet, daily life, medicine, and industry have all been inseparable from nitrates. Nitrates are widely used as food additives, fertilizers for crops, oxidizing agents in steelmaking, heat treatment agents for aluminum alloys, and raw materials for explosives. The medicinal use of nitrates can be traced back more than 2000 years, when potassium nitrate (KNO_3, also known as saltpeter) was used to treat abscesses and oral ulcers, as depicted in Dunhuang murals. Nitroglycerin, a nitrate-based drug, was initially employed by Alfred Nobel, the "Father of Explosives," as a fundamental material for cannons and firearms. Later, however, it was discovered that nitroglycerin also dilates blood vessels. Transforming an explosive into a therapeutic drug, Nobel made a remarkable contribution. To this day, nitroglycerin is still used sublingually in medicine to relieve angina pectoris.

By the twentieth century, nitrates in food had long been regarded as harmful, primarily because they were thought to be reduced in the body to nitrites, which in turn contribute to the formation of nitrosamines strongly associated with cancer. As a result, strict limits were imposed on nitrate levels in food and drinking water, and the World Health Organization recommended restricting daily nitrate intake. However, with advances in scientific research, growing evidence has shown that nitrates and nitrites can be beneficial to health, being both stable and safe—entirely different from the harmful nitrosamines. In 2010, the International Agency for Research on Cancer, and in 2017, the World Health Organization issued official statements based on extensive clinical and basic research, concluding that there is no substantial evidence supporting the carcinogenicity of nitrates or nitrites.

Our research group's interest in nitrates began in the 1980s, when the author was pursuing a doctoral degree at Beijing Medical University (now Peking University Health Science Center). Literature at the time reported that the concentration of nitrate in human saliva is extremely high—five to ten times that in blood—yet most scholars then considered nitrates and nitrites harmful. The author was puzzled:

could such a high concentration of nitrate in saliva really harm the body itself? With this question in mind, and later as an independent mentor at the Beijing Stomatological Hospital, Capital Medical University, the author finally had the opportunity to begin a long-term exploration of this field. Our first question was whether the high concentration of salivary nitrate is physiological or pathological. If such high levels exist in normal individuals, they must be physiological; human evolution would not evolve a mechanism that harms itself. Simple logical reasoning suggested that common perceptions about nitrates might be mistaken. Thus, graduate students were tasked with recruiting healthy volunteers to measure nitrate concentrations in parotid saliva, mixed saliva, and blood from the same individuals, and further verified in healthy miniature pigs that high salivary nitrate is indeed physiological. From there, we posed a series of scientific questions and gradually carried out systematic investigations: Which salivary gland is primarily responsible for nitrate secretion? Do salivary gland cells have a nitrate transporter channel? What roles does high salivary nitrate serve? How is nitrate metabolized in the body? What systemic functions does it exert? And is nitrate truly safe? Through collaborative efforts, we conducted systematic research into these questions and discovered that nitrate plays a crucial role in maintaining homeostasis and protecting bodily functions via the nitrate–nitrite–nitric oxide (NO) pathway and the cell membrane nitrate transporter (Sialin). Ultimately, based on this body of work, we proposed the novel concept of "homeostatic medicine" and began developing a nitrate-based drug named "Nitrator." After more than two decades, these scientific questions have finally yielded a relatively complete answer.

Our passion for scientific research has fueled our determination and desire to explore the unknown world, though it has often been accompanied by persistent feelings of helplessness and melancholy. During scientific inquiry, we have encountered countless setbacks, confusions, moments of helplessness, and even despair. Yet time and again, we have overcome them with resilience and perseverance. In 2022, after I had delivered an academic lecture titled "Nitrate: From the Oral Cavity to a Guardian of Systemic Health," Professor Du Xian, then Editor-in-Chief of the People's Medical Publishing House, who attended the entire talk, extended a sincere invitation for our team to write a monograph on nitrates. By such a serendipitous opportunity, the book *Inorganic Nitrate and Human Body Homeostasis* has come into being. This monograph records the years our team has spent exploring the subject of nitrates and recounts many unforgettable stories along the way. The content and writing of this book may not be especially remarkable, but they are an authentic reflection of our scientific exploration and cherished memories. Throughout this long and challenging journey of research, we have received tremendous support and assistance from many teachers, experts, colleagues, friends, science and technology management agencies, and our families. These include my doctoral advisor, Professor Zhaoju Zou of Peking University School of Stomatology; Professor Dajun Deng of Peking University Cancer Hospital, with whom I have conducted joint studies; Dr. Xibao Liu and Dr. Indu S. Ambudkar of the US National Institutes of Health; Professor Jon O. Lundberg of the Karolinska Institute in Sweden; and, of course, our own research group, full of innovation, unity, and strength. Here, I would like to extend my most sincere and heartfelt gratitude to them all.

The beauty of life, the beauty of nature, and the beauty of science—I hope that every teacher and friend who opens this book can share these with us and together experience the truth and beauty of exploring the halls of science, both distant and near.

Beijing, China Songlin Wang

Contents

Part I

The Debate Over Inorganic Nitrate

Nitrates are a kind of inorganic salt commonly found in nature. They are found in food and drinking water, with higher levels in beets, spinach, celery, and other green vegetables. Excessive nitrites and nitrosamines in food can cause poisoning, cancer, or even death. However, it remains controversial as to whether nitrates affect body health. With the development of scientific research in recent years, the beneficial effects of nitrates on the human body have been gradually confirmed, leading to an increase in their medicinal value. But what exactly are nitrates? Throughout history, both our understanding of nitrates and the way we think about them have evolved. What are the potential applications of nitrates? This chapter will explore each of these questions in detail.

History of Inorganic Nitrates

1

Wen Pan, Li Chi, Jiabo Wang, and Yuanchun Si

Since ancient times, nitrates have been indispensable to human diet, life, medicine, and industry. They serve as food additives, including pickling salt and hair color agents, and can play a role in seasoning. They can also be used as fertilizer for crops, especially root crops such as beets and radishes, and as an accelerant and heat treatment agent for steelmaking, in aluminum alloy, and in the production of explosives. Traceable written records indicate that the medical application of nitrates in China has a history of more than 2000 years.

1.1 What Are Nitrates?

Wen Pan

1.1.1 The Definition of Nitrates

Nitrates are ionic compounds, generally referred to as metal ions or salts composed of ammonium ions and nitrate ions. Common nitrates are sodium nitrate, potassium nitrate, ammonium nitrate, calcium nitrate, lead nitrate, cerium nitrate, and so on. Almost all nitrates are soluble in water. Only urea nitrate is slightly soluble in water, while bismuth nitrate is insoluble in water.

W. Pan · Y. Si
Beijing Stomatological Hospital, Capital Medical University, Beijing, China

L. Chi · J. Wang (✉)
School of Traditional Chinese Medicine, Capital Medical University, Beijing, China
e-mail: jiabo_wang@ccmu.edu.cn

S. Wang (ed.), *Inorganic Nitrate and Human Body Homeostasis*,
https://doi.org/10.1007/978-981-95-2824-0_1

1.1.2 The Circulation of Nitrates in Nature

In nature, nitrogen exists in three forms: molecular nitrogen, inorganic nitrogen, and organic nitrogen. The process of the nitrogen cycle in nature is one of the basic material cycles in the biosphere. The air contains about 78% nitrogen, and the nitrogen in the atmosphere enters the soil through the actions of microorganisms and other processes. It is used by animals and plants and finally returns to the atmosphere with the participation of microorganisms (Fig. 1.1). Nitrates are abundant in nature. The main source of nitrates is nitrogen fixation by nitrogen-fixing bacteria. When exposed to the high temperature of lightning, nitrogen and oxygen in the air combine directly to form nitrogen oxides, which dissolve in rain to form nitric acid. Nitric acid reacts with minerals on the ground to form nitrates.

Nitrates are the main source of nitrogen obtained by land plants from the soil and are indispensable for maintaining plant growth. Plants absorb nitrates from the soil through the proton/nitrate coupling mechanism, which are then transported to leaves for storage and/or further assimilation. Nitrates are used to synthesize chlorophyll to promote photosynthesis, and to synthesize amino acids to further synthesize various proteins.

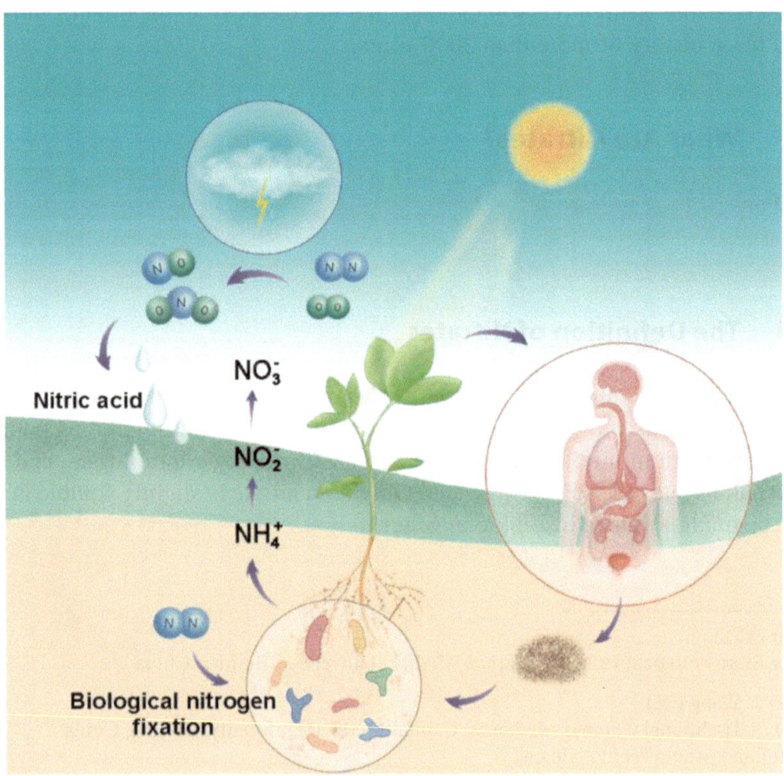

Fig. 1.1 Schematic diagram of the natural nitrogen cycle

Animals directly or indirectly eat organic nitrogen (proteins) or inorganic nitrogen (nitrates) synthesized by plants from the food chain, which are then broken down and used by their own bodies. In the metabolic process of animals, part of the protein is decomposed into ammonia, uric acid, and urea and discharged into the soil. The organic nitrogen in the remains of plants and animals is converted by microorganisms into inorganic forms (ammonia and nitrate), thus completing the nitrogen cycle of the ecosystem. Therefore, nitrates are widely found in human drinking water and food, especially in green leafy vegetables.

1.2 Application of Nitrates in Ancient Chinese Medicine

Li Chi and Jiabo Wang

The main component of saltpeter (Fig. 1.2) is potassium nitrate (KNO_3). In traditional Chinese medicine (TCM), this compound was historically designated by multiple nomenclature variants, including "消石" (*Xiao Shi*; niter), "火硝" (*Huo Xiao*; lit. "fire niter"), "焰硝" (*Yan Xiao*; lit. "flame niter"), and "化金石" (*Hua Jin Shi*; lit. "gold-dissolving stone"), as documented in classical pharmacopeias [1]. Historical records indicate that the history of the medical application of saltpeter in China spans more than 2000 years [2].

The term "*Xiao Shi*" was first used in the "*Prescriptions for Fifty-Two Diseases*" (*Wu Shi Er Bing Fang*, 五十二病方) unearthed from the Mawangdui Han Dynasty

Fig. 1.2 Saltpeter

tombs (168 BCE) in Changsha city of Hunan Province, China [3], which is generally regarded by historians as having been finished in the Warring States period (475–221 BCE) or even earlier [4]. There is a therapeutic protocol recorded in the medical text involving "immersing *Xiao Shi* in warm water for the treatment of carbuncles" (Fig. 1.3), meaning that saltpeter (niter) was dissolved in warm water and then used externally to treat carbuncles. It can be seen that the sages of ancient China used saltpeter topically as early as during the Warring States period for surgical treatment of diseases. This could be the world's earliest record of the medical application of saltpeter.

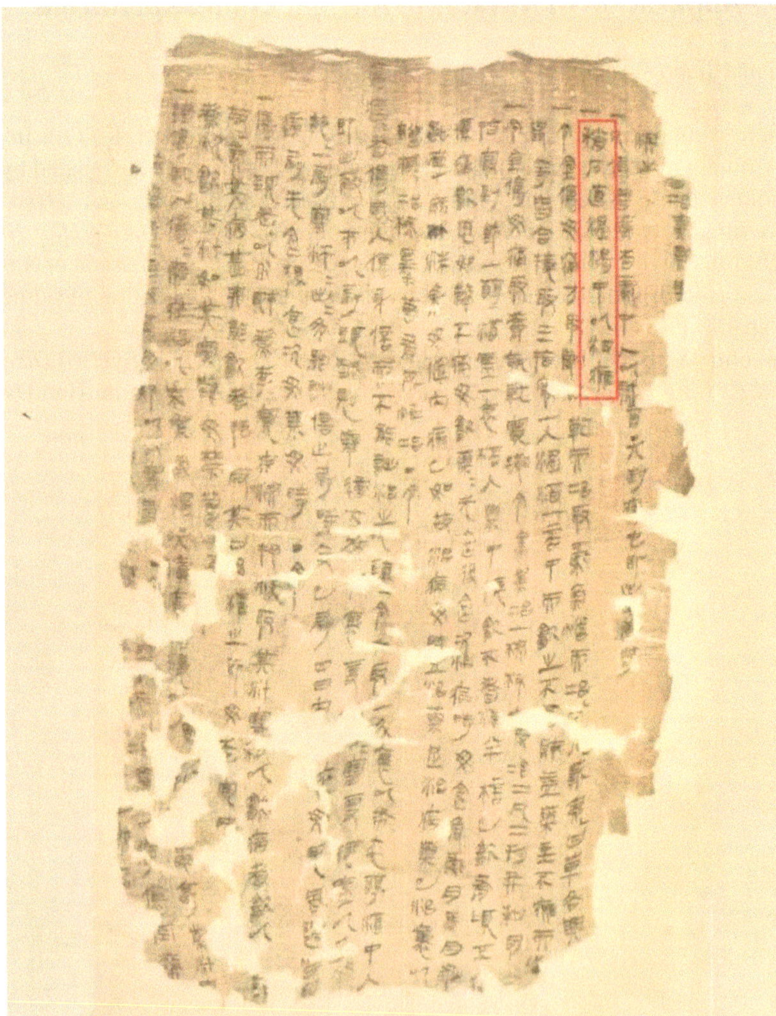

Fig. 1.3 *52* A prescription, *Prescriptions for Fifty-Two Diseases*, Hunan Provincial Museum Collection

A medical case involving Chunyu Yi (淳于意, 215–140 BCE), a renowned physician of the Western Han Dynasty, was recorded in the "*Biography of Bian Que and Cang Gong*" (*Bian Que Cang Gong Lie Zhuan*, 扁鹊仓公列传), a chapter of "*Records of the Grand Historian*" (*Shi Ji*, 史记) [5]. In treating Wang Meiren (王美人) of Zichuan County, manifesting "restless pulse" (*Zao Mai*, 躁脉; clinically interpreted as blood-heat agitation), Chunyu Yi prescribed *Xiao Shi*. The text records: "The restless pulse indicates presence of residual conditions. Administer one dose of *Xiao Shi*, followed by bloodletting equivalent to five or six bean volumes" [6], which may be the earliest known medical case of saltpeter being taken orally to treat a disease.

Due to the limitations of the ancients in understanding of substances, there was some confusion when referring to saltpeter (KNO_3) and mirabilite (Na_2SO_4) [7]. However, the Herbalism monograph "*Shennong's Classic of Materia Medica*" (*Shennong Bencao Jing*, 神农百草经*)*, which was written before the Eastern Han Dynasty (25–220 CE), has recorded that they were actually different substances. The book recorded that saltpeter (*Xiao Shi*) governs "the five *zang (*脏*)* organ heat, stomach's being open and closed, washing away the accumulation of food and fluids, discharging the old, and absorbing new *qi* (气*, a concept in traditional Chinese medicine*) and discharging pathogenic *qi,*" whereas mirabilitum governs "all diseases, discharging cold and heat pathogenic *qi*, and expelling the accumulation and solidification of the six bowels" and can "eliminate seventy-two kinds of calculus." It can be seen that saltpeter and mirabilitum could be taken internally back then, and the scope of disease treatment is also wider than today. In addition, "*Shennong's Classic of Materia Medica*" classified both saltpeter and mirabilitum as "top grade" (Fig. 1.4). The description of "top grade" drugs in the book was that they "govern nourishing life to conform to the heavens, non-toxic; excessive and prolonged use does not harm the body,"indicating that the ancients believed that oral saltpeter was relatively safe, which may be the world's earliest record of the safety of saltpeter being taken orally [2].

The book of heirloom and unearthed documents, formed around the same period of "*Shennong's Classic of Materia Medica,*" also recorded that saltpeter was used in medical practice, such as "*Essential Prescriptions from the Golden Cabinet*" (*Jin Gui Yao lüe*, 金匮要略*)*, formed in the Eastern Han Dynasty, recorded "*Da Huang Xiao Shi Tang"* (大黄硝石汤) for treatment of jaundice, and "*Xiao Shi Fan Shi San"* (硝石矾石散) for treatment of black jaundice for women in labor [8]. "*Wuwei Han Bamboo Slips*" (*Wu Wei Han Jian*, 武威汉简) formed before the Eastern Han Dynasty, recorded that saltpeter taken orally and typically used to treat surgical diseases [9]. Dunhuang volume "*Zhang Zhongjing's Five Viscera Theory*" (*Zhang Zhongjing Wu Zang Lun*, 张仲景五脏论) formed in the Jin Dynasty and Tang Dynasty, recorded saltpeter being used to treat carbuncles and swellings [10].

It is worth mentioning that in Taoist works, there are many records of saltpeter being taken orally or made into elixirs. "*The Biography of the Immortals*" *(Lie Xian Zhuan*, 列仙传), likely written during the Western Han Dynasty (206 BCE–8 CE), records that the immortal Chi Fu (赤斧; lit. "red axe") used saltpeter as a companion for taking elixirs, and that he "could make infinite amounts of water, refine elixirs,

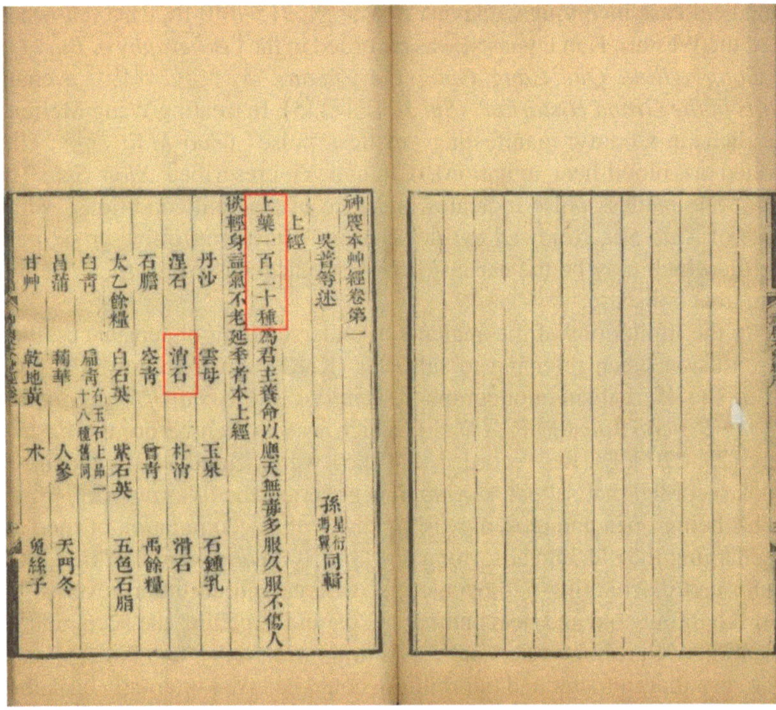

Fig. 1.4 *Shennong's Classic of Materia Medica*, Volume 1, edited by Sun Xingyan and Sun Fengyi, National Library of China Collection

and take them with saltpeter." The book "*Thirty-Six Shui Fa*" *(San Shi Liu Shui Fa,* 三十六水法*)* recorded 46 kinds of "*Shui Fa*" (水法; lit. "water method," a concept in Taoism) for alchemy, including 59 formulas, 32 of which used saltpeter, such as in the preparation method of "andrographis water," which involved "taking 1 *jin (*斤*, a traditional Chinese unit of mass)* of realgar, putting in the bamboo tube, adding 4 *liang (*两*, a traditional Chinese unit of mass)* of saltpeter, sealing the pipe with paint, putting it in the pool, and the water would be generated 30 days later" [7]. Alchemists first recognized saltpeter was combustible, so refining elixir with saltpeter often used the "*Shui Fa*" or volatile fire method [11]. During the Liang Dynasty (502–557 CE) of the Northern and Southern Dynasties period (420–589 CE), alchemist and medical expert Tao Hongjing (陶弘景) discovered the flame color test saltpeter method, which involved: "When burned, it emits purple and green smoke, and it turns into ash afterward" [12]. Alchemist's correct understanding of the chemical properties of saltpeter gave rise to the invention of gunpowder, and distinguished saltpeter from other mineral medicines, such as mirabilitum.

On the basis of the alchemist's understanding, medical scientists in the Song Dynasty (960–1279 CE) essentially understood saltpeter. Therefore, since the Song Dynasty, the medical application of saltpeter became more and more extensive, and gradually expanded the application of internal medicine, surgery, gynecology, ophthalmology, and other disease treatments. The relevant records are numerous, and these will not be discussed here. Interested readers can read the article "Efficacy Research of Niter Based on Ancient Medical Books" [13].

Based on the ancient medical applications of saltpeter, its effects can be broadly classified into seven categories: pain relief, detoxification and swelling, removing decay and regenerating muscle, softening and dispersing knots, opening the orifices and removing filth, inducing dampness and remove jaundice, and warming the interior and activating the blood. These effects could be applied in the treatment of internal medicine, surgery, gynecology, ophthalmology, and other illnesses (Table 1.1).

Ancient literature records that people used saltpeter to treat oral diseases at a very early stage, such as "*Ri Hua Zi's Materia Medica*" (*Ri Hua Zi Ben Cao*, 日华子本草), in which it was recorded that "ablative stone … contained in the mouth to cure the throat closure" [14]. The book is considered to be written in the period of the Northern Qi Dynasty (550–577 CE), the Kaiyuan era of the Tang Dynasty (713–741 CE), the Five Dynasties and the Ten Kingdoms (Wuyue Kingdom, 907–978 CE), and the Kaibao era of the Northern Song Dynasty (968–976 CE) [15]; thus, the use of saltpeter for oral diseases can be traced back at least to the Kaibao era. Another herbal book of the Northern Song Dynasty (960–1127 CE), "*Zheng Lei Bencao*" (证类本草), also recorded that "saltpeter" was used for "mouth sores, paralysis of the throat and pharyngeal congestion, and tooth and jaw swelling

Table 1.1 Basic information on the efficacy and application of saltpeter recorded in ancient literature

Efficacy	Applicable syndrome	Usage
Pain relief	Headaches, toothaches, pain in other parts of the body	Mostly as powder, applied to the affected area or mucous membrane parts such as the inside of the nose and the corners of the eyes, or as a decoction for oral administration
Detoxification and swelling	Throat paralysis, mouth sores, eye diseases	Mostly as powder, applied to the affected area or mucous membranes such as the inside of the nose and the corners of the eyes
Removing decay and regenerating muscle	Canker sore	Mostly used topically as tonics, powders, or elixirs
Softening and dispersing knots	Scrofula, concretions, and conglomerations	Mostly taken orally as pills or applied topically to the lesion
Opening the orifices and removing filth	Heat and cold, fainting of the mind, and closing of the orifices	Mostly taken orally as pills or applied to the mucous membranes inside the nose and the corners of the eyes with powder
Inducing dampness and remove jaundice	Jaundice, strangury	Taken orally as a decoction or powder
Warming the interior and activating the blood	Yin-cold excess pattern, cold, and blood stasis	Taken orally as pills or decoction

and pain" [16], indicating that saltpeter was used for the treatment of mouth sores, paralysis of the throat, and toothache during the Kaibao era of the Northern Song Dynasty. Since then, saltpeter is widely used in the treatment of oral diseases such as pharyngeal diseases, toothache, mouth sores, heavy tongue, and dental chancre, etc., which can be found in the article "Ancient application of nitrates in oral diseases—Research progress of nitrates in Stomatology" [17].

Saltpeter is mostly applied locally in powder for the treatment of oral diseases. Taking the treatment of mouth sores as an example, saltpeter can be applied alone by "sprinkling saltpeter powder on the wound surface three or five times a day" for the treatment of children's mouth sores [18], and can be used with succus bambusae, borax, and other clear-heat and phlegm-dissolving traditional Chinese medicine [19, 20], but also with pollen typhae, indigo naturalis, borax, licorice, etc., composed of formulas, such as "*Bi Xue*" and other topical applications (Table 1.2) [21].

To sum up, in ancient China, saltpeter offered a wide range of treatments for diseases, with diversified dosage forms and routes of administration, which could be taken orally for systemic regulation by soups, pills, as well as locally applied by pills, oranas, which embodied the characteristics and advantages of traditional Chinese medicine treatments. The rapid effect of saltpeter in the treatment of pain, paralysis of the throat, and other diseases [5–7] can also be taken as a typical representative supporting "the effectiveness of traditional Chinese medicine in treating diseases is not slow."

1.3 From Explosive to Medicine: Nitroglycerin—Contributions by Mr. Nobel

Yuanchun Si

The clinical application of nitroglycerin came approximately 2000 years after that of saltpeter. This delay was not only due to nitroglycerin being synthesized just over 170 years ago but also because initial reports focused on its potent explosive capabilities. Through continuous improvements by Mr. Nobel, the explosive properties of nitroglycerin were eventually made safe and controllable. Nobel's successful commercialization of nitroglycerin facilitated broader access to this compound, laying the material foundation for its medical applications (Fig. 1.6).

1.3.1 Physical and Chemical Properties of Nitroglycerin

Nitroglycerin ($C_3H_5N_3O_9$), also known as glyceryl trinitrate or trinitroglycerin, is a colorless to pale yellow oily liquid with a molecular weight of 227.08 (Fig. 1.7). It has limited solubility in water and poses a risk of explosion when subjected to extreme cold or heat, impact, friction, or exposure to open flames or high temperatures. In medicine, nitroglycerin serves as a vasodilator used in the treatment of angina pectoris. Its clinical administration includes sublingual administration, inhalation, intravenous infusion, oral intake, and transdermal delivery.

Table 1.2 Usage of saltpeter in the treatment of mouth sores in canonical texts of different historical periods

Dynasty	Source	Formula	Composition and usage	Efficacy and indications
Song dynasty (960–1279 CE)	*Formulas from Benevolent Sages Compiled during the Taiping Era*	Realgar powder formula	Realgar, saltpeter, felserpentis, radix sophorae flavescentis, cinnabaris, fibroferrite, and moschus	Treating mouth sores in children which are painful and rotting, regardless of whether the sores are white or red, or between the cheeks and jaws, or on the gums. This formula is recommended for children with prolonged mouth sores [22]
Song dynasty (960–1279 CE)	*Formulas to Aid the Living*	Green snow	Pollen typhae, indigo naturalis, borax, saltpeter, licorice (each in equal parts)	Treating all kinds of heat, swollen throat, inability to swallow, sores in the mouth and tongue, tightness and strength at the root of the tongue, incorrect speech, swelling, and pain in the cheeks [19]
Song dynasty (960–1279 CE)	*Comprehensive Recording of Divine Assistance*	Mirabilitum powder formular	Mirabilitum, saltpeter, chalcanthitum, alumen, natrii sulfas (all five flavours withered and dried), glauberitum (burnt), bombyx batryticatus (straight fried), radix et rhizoma glycyrrhizae, indigo naturalis (each in equal parts)	Or mouth sores and throat closure [23] (Fig. 1.5)
Ming dynasty (1368–1644 CE)	*Formulas for Universal Relief*	/	Succus bambusae and saltpeter and apply on affected part	HYeavy tongue, goose mouth [19]
Ming dynasty (1368–1644 CE)	*Simple Formulas for Health*	Another formula for treating mouth sores in children	Saltpeter powder on the sores, three to five times a day	Mouth sores in children [18]

(continued)

Table 1.2 (continued)

Dynasty	Source	Formula	Composition and usage	Efficacy and indications
Ming dynasty (1368–1644 CE)	*Orthodoxy of Medicine*	Another prescription for mouth sores	Saltpeter and borax, borax held in the mouth without opening; arisaematis powder for the end of vinegar, paste on yǒng quán on the sole of the foot	Mouth sores [20]
Qing dynasty (1644–1911 CE)	*First Aid Collection for Sentient Beings*	/	Rhizoma coptidis, cortex phellodendri chinensis (powder, eight points), realgar, indigo naturalis, saltpeter (two points), calculus bovis artifactus, borneolum, borax, cinnabaris (one point), all grinding into powder, each with a little, blowing into the mouth to cure within a short time	Sores on the mouth, inability to suckle on a heavy palate and swollen, and blocked throat, and all kinds of heat and poison [23]

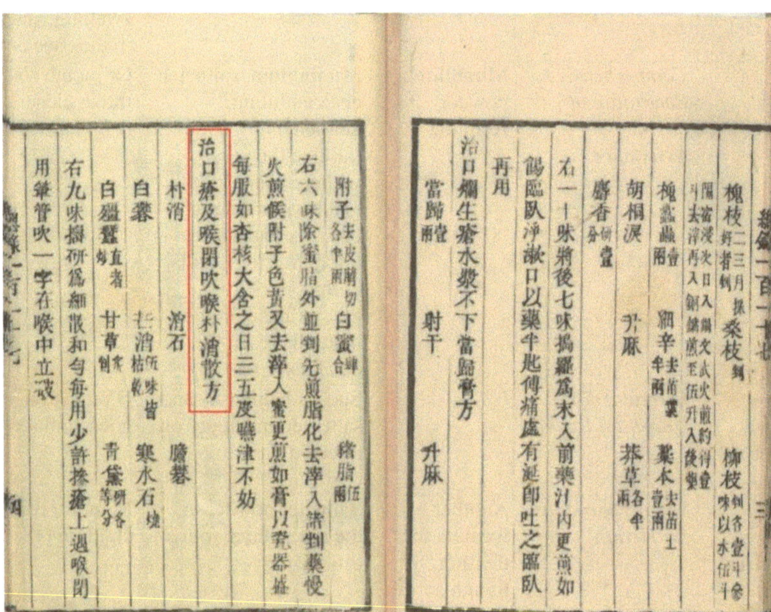
槐枝二三月採好者剉 桑枝剉 柳枝剉各壹斗以水伍斗
隔宿浸次日入鍋文武火煎約得壹斗去滓再入銅鍋慢火煎至伍升入後藥
槐蠹蟲壹兩 細辛去苗葉半兩 藁本去苗土壹兩
胡桐淚 升麻 莽草各半兩
麝香研壹分
右一十味將後七味搗羅為末入前藥汁內更煎如
餳臨臥淨漱口以藥半匙傅痛處有涎即吐之臨臥
再用
治口爛生瘡水漿不下當歸膏方
當歸壹兩 射干 升麻
附子去皮臍各半兩切 白蜜肆合 豬脂伍兩
右六味除蜜脂外並剉先煎脂化去滓入諸剉藥慢
火煎候附子色黃又去滓入蜜更煎如膏以瓷器盛
每服如杏核大含之日三五度嚥津不妨
治口瘡及喉閉吹喉朴消散方
朴消 滑石 膽礬
白礬 寒水石燒
白殭蠶直者炒 甘草炙剉 青黛各研等分
右九味搗研為細散和勻每用少許摻瘡上遇喉閉
用筆管吹一字在喉中立效

Fig. 1.5 *General Records of Holy Universal Relief (Sheng Ji Zong Lu,* 圣济总录) Volume 117, National Library of China Collection

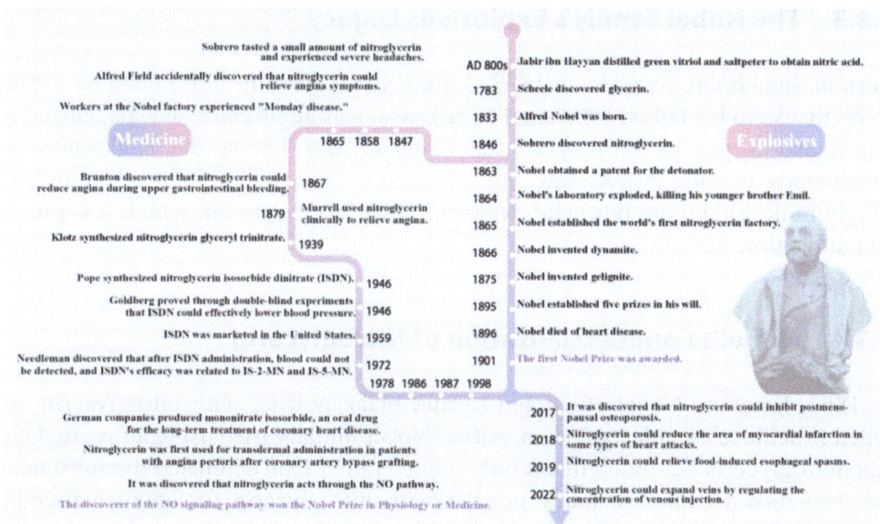

Fig. 1.6 Chronology of nitroglycerin from explosive to medicine

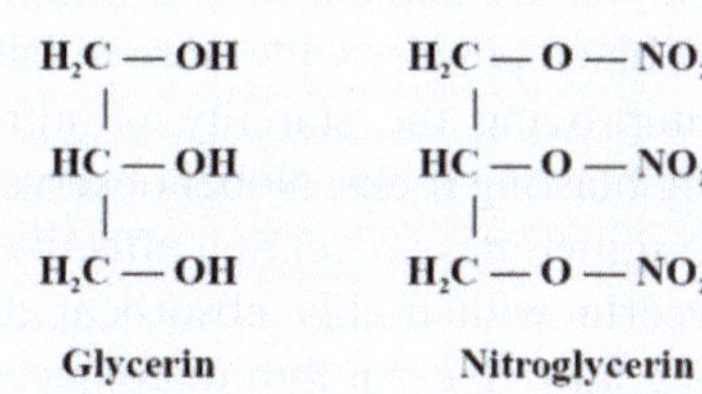

Fig. 1.7 Molecular formulas of glycerol and nitroglycerin (the latter has three NO_2 groups substituting for the three H atoms in glycerol) [24]

1.3.2 Discovery of Nitroglycerin

In the eighth century, Arab alchemist Jabir ibn Hayyan discovered nitric acid while dry distilling green vitriol (iron(II) sulfate heptahydrate) and saltpeter (potassium nitrate). In 1783, pharmacist Carl Wilhelm Scheele identified "sweet oil" (glycerol) during his experiments with olive oil and lead oxide. Fast forward to 1846, Italian chemist Ascanio Sobrero synthesized nitroglycerin by nitrating glycerol with a mixture of nitric and sulfuric acids, producing a yellow nitro compound. This reaction is highly exothermic and can easily lead to explosions if not cooled promptly [24].

Sobrero demonstrated nitroglycerin's explosive properties at a lecture at Turin University in 1847, describing its taste as sweet, pungent, and aromatic [24]. He noted that even a small drop on the tongue could cause severe headaches lasting for hours. An explosion during one of his experiments left him with serious facial scars. Reflecting on the numerous casualties caused by nitroglycerin accidents, Sobrero expressed deep regret over having been its discoverer [25, 27].

1.3.3 The Nobel Family's Explosives Legacy

Born in Stockholm, Sweden, in 1833, Alfred Nobel grew up surrounded by explosives, thanks to his father, Immanuel Nobel, who was an inventor and specialized in mines and torpedoes [26]. After the Crimean War led to the bankruptcy of Immanuel's factory, Alfred saw the commercial potential of nitroglycerin and sought to develop more powerful explosives than black powder, which was prevalent at the time [28, 29].

1.3.4 Nobel's Commercialization of Nitroglycerin

In 1862, Emanuel Nobel attempted to mix black powder with nitroglycerin but failed to achieve a stable explosion. Alfred Nobel initially tried using a fuse to detonate nitroglycerin but found that it only ignited rather than detonated the substance. He then mixed explosive black powder with nitroglycerin for another attempt. However, the resulting mixture was unstable and, after several hours, would only burn rather than explode.

Nobel then experimented with using a fuse to first ignite black powder, utilizing the pressure from its explosion to trigger the detonation of nitroglycerin. This approach yielded stable results. Building on this idea, Nobel replaced the moisture-sensitive black powder with more reliable and powerful mercury fulminate, significantly improving the stability of nitroglycerin detonation. During experiments involving blasting rocks, Nobel observed that nitroglycerin could leak through rock fissures, sometimes preventing effective detonation. To address this issue, he mixed nitroglycerin with highly absorbent diatomaceous earth, creating dynamite. By adsorbing nitroglycerin into diatomaceous earth, he greatly enhanced its transport stability and practicality.

After large-scale industrial production of dynamite, Nobel sought to improve its explosive power and address its sensitivity to moisture. Through extensive research, he discovered nitrocellulose, which has excellent absorption and explosive properties. Nobel combined nitroglycerin with nitrocellulose to create gelignite. Unlike dynamite, gelignite does not contain diatomaceous earth, preventing nitroglycerin from leaking due to moisture. The hydrophobic nature of nitrocellulose allowed gelignite to be detonated underwater, broadening its applications.

Initially, nitroglycerin was poured directly into drilled rock holes. Subsequent improvements involved packaging mixtures of nitroglycerin with black powder, diatomaceous earth, or nitrocellulose in cardboard boxes to enhance stability and explosive power (Fig. 1.8). These mixtures needed to be prepared immediately before use; if left for several hours, the nitroglycerin would fully absorb into the spaces within the explosive material, rendering it non-detonable. Nobel's innovative approach involved first detonating the inner layer of black powder, using its explosive energy to trigger the outer layer of nitroglycerin.

Repeated experiments confirmed the feasibility of this approach. Nobel later replaced black powder with the more effective mercury fulminate, which generated

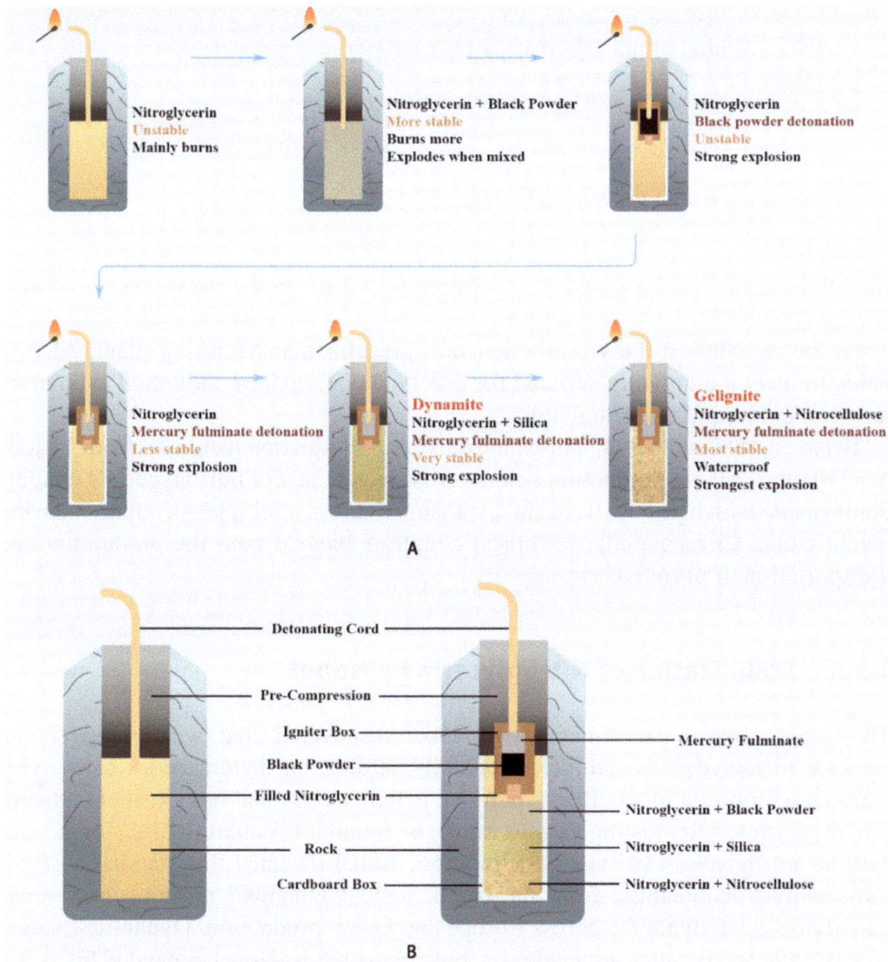

Fig. 1.8 Alfred Nobel's development process for explosives. (**a**) Alfred Nobel experimenting with mixtures of black powder and nitroglycerin. (**b**) Schematic diagrams of early explosives, dynamite, and gelignite

stronger gas production and thus greater explosive power (Fig. 1.9). In 1863, Nobel obtained a patent for his blasting cap device [26]. The immense power of nitroglycerin brought Nobel a flood of orders.

However, the development process was fraught with challenges. On September 3, 1864, an explosion at the laboratory resulted in the death of his younger brother Emil Nobel and several others. The explosion destroyed the laboratory entirely, and local authorities subsequently banned Nobel from conducting experiments within city limits. Forced to relocate, Nobel moved his laboratory to a barge on Lake Mälaren. Due to the instability and safety concerns associated with experimentation on the barge, in 1865, he relocated again to an uninhabited area near Lake Mälaren,

Black Powder: $2KNO_3+S+3C=K_2S+N_2\uparrow+3CO_2\uparrow$

Nitroglycerin: $4C_3H_5N_3O_9=6N_2\uparrow+12CO_2\uparrow+O_2\uparrow+10H_2O\uparrow$

Nitrocellulose: $2(C_6H_7O_{11}N_3)_n=3nN_2\uparrow+7nH_2O\uparrow+3nCO_2\uparrow+9nCO\uparrow$

Mercury fulminate: $Hg(CNO)_2=Hg+2CO\uparrow+N_2\uparrow$

Fig. 1.9 Explosive reaction equations for components used by Alfred Nobel during his research

where he established the world's first nitroglycerin manufacturing plant. Also in 1865, to meet the growing demand for orders, Nobel opened additional explosives factories in Finland and Hamburg.

Between 1867 and 1874, annual nitroglycerin production increased from 11 tons to 3210 tons [30]. As production volumes rose, awareness of nitroglycerin's dangers diminished, which led to frequent explosions across multiple locations, causing public panic. Consequently, European countries banned both the production and transportation of nitroglycerin.

1.3.5 Stabilization of Nitroglycerin by Nobel

Through extensive experimentation, Nobel discovered that wood chips could enhance nitroglycerin's stability, eventually settling on diatomaceous earth. This material, composed of fossilized diatoms, is lightweight and has excellent adsorption properties. After testing various ratios, he found a 1:3 mixture of diatomaceous earth to nitroglycerin worked best. In 1866, Nobel invented this stabilized explosive, naming it dynamite from the Greek word "dynamis," meaning power. He quickly patented dynamite across Europe and began production. Dynamite's safety against falls and hammering made it popular worldwide. Nobel expanded his factories across Sweden, Norway, and Finland. The commercial success led to Nobel and his father receiving the Légion d'honneur gold medal in 1868. Dynamite significantly advanced mining and transportation industries and played a crucial role in Prussian victory during the Franco-Prussian War [31]. The rapid development of railway projects and mines across Europe fueled the explosive growth of Nobel's factories.

1.3.6 Enhancing Nitroglycerin's Explosive Power

Nobel sought to increase dynamite's explosive potential. While diatomaceous earth absorbed large amounts of nitroglycerin, it did not contribute to the explosion, reducing overall potency. Through extensive research, Nobel replaced diatomaceous earth with nitrocellulose, finding an optimal mix of 92% nitroglycerin and 8%

nitrocellulose. This paste-like substance retained nitroglycerin's explosive power while maintaining dynamite's relative safety. Named gelignite, this new explosive was non-reactive to impact or water and safe for long-distance transport [32, 33].

1.3.7 Discovery of Nitroglycerin's Medical Value

In 1844, Antoine Balard from the University of Paris reported severe headaches caused by inhaling amyl nitrite vapors. In 1847, Ascanio Sobrero experienced intense headaches lasting several hours after tasting a tiny amount of nitroglycerin on his tongue, deterring him from further personal experimentation with this volatile substance.

In 1858, British doctor Field observed that a 67-year-old female patient's chest pain was alleviated by nitroglycerin. However, due to its side effect of causing severe headaches, few were willing to use it.

By 1865, shortly after Nobel's factory opened, workers exhibited a peculiar phenomenon known as "Monday disease," characterized by headaches upon returning to work, which gradually subsided over time. Those who changed out of their uniforms before going home did not experience these headaches. Interestingly, workers with angina pectoris reported reduced symptoms at work. Further investigation revealed that workers absorbed nitroglycerin through their skin, leading to vasodilation and symptom relief [34].

At the time, medical practices included controversial homeopathy, which proposed treating diseases with substances that produced similar symptoms in healthy individuals—essentially using headache-inducing drugs to treat headaches. This principle led to various experiments exploring nitroglycerin's therapeutic potential. In 1867, Lander Brunton published in *The Lancet* that amyl nitrite could treat angina during upper gastrointestinal bleeding [35].

It wasn't until 1878 that William Murrell experimented with diluted nitroglycerin for treating angina and lowering blood pressure. In 1879, Murrell published a paper in *The Lancet* detailing the method of using nitroglycerin for angina, marking the beginning of its widespread clinical application [36].

Nitroglycerin proved to have a longer-lasting therapeutic effect compared to amyl nitrite, was easier to use, and provided stable efficacy, accelerating its market adoption. In 1940, Krantz synthesized mannitol nitrate, which offered more prolonged vasodilation than nitroglycerin. In 1946, Porje in Sweden developed isosorbide dinitrate (ISDN). He noted that ISDN had higher toxicity but could control angina pectoris throughout the day by reducing oral doses (40 mg per dose, three times daily). That same year, Goldberg and colleagues conducted the first double-blind study, discovering that ISDN, also known as Angicet, effectively lowered blood pressure for extended periods. ISDN became available in the U.S. market in 1959.

In 1972, American researcher Philipp Needleman found that ISDN's parent compound disappeared after oral administration. He concluded that it lost its effectiveness after liver metabolism, sparking significant interest. Clinical studies confirmed

ISDN's substantial clinical benefits. To address the paradox of ISDN's disappearing parent compound, manufacturers conducted extensive research. They discovered that ISDN's effects were primarily due to its metabolites, 2-mononitrate and 5-mononitrate of isosorbide. Further research led to the development of 5-mononitrate of isosorbide, better suited for long-term coronary heart disease treatment, with no first-pass liver elimination and a utilization rate of up to 100%, far exceeding ISDN's 30% and nitroglycerin's 8%.

By 1986, nitroglycerin was first used transdermally for ischemic reperfusion in flap transplantation. In 1987, it was discovered that nitrate drugs act via nitric oxide (NO). Studies in the 1970s revealed that nitroglycerin metabolizes into NO in the bloodstream, leading to vasodilation through cyclic guanosine monophosphate (cGMP). This clarified that the unique biological effects of "nitrates", used for over a century, were mediated by NO, a gas signaling molecule present in almost all organisms [37–39].

Currently, NO relieves angina primarily through two mechanisms (Fig. 1.10):

1. *Smooth muscle relaxation*: NO acts on smooth muscle cells, reducing myosin phosphorylation through cGMP, leading to vasodilation. The vasodilatory effects of nitrates are dose-dependent: small doses dilate capacitance vessels, moderate doses dilate coronary arteries, and large doses dilate small arteries.
2. *Platelet aggregation inhibition*: NO increases cGMP levels in platelets, inhibiting the binding of fibrinogen to glycoprotein IIb/IIIa receptors, which are crucial for platelet aggregation. This mechanism is also considered key to NO's enhancement of coronary perfusion.

In 2005, Austrian researchers discovered that nitroglycerin's effects are mediated through mitochondrial aldehyde dehydrogenase 2 (ALDH2). Individuals lacking ALDH2 experience facial flushing after alcohol consumption and may have a poorer response to sublingual nitroglycerin.

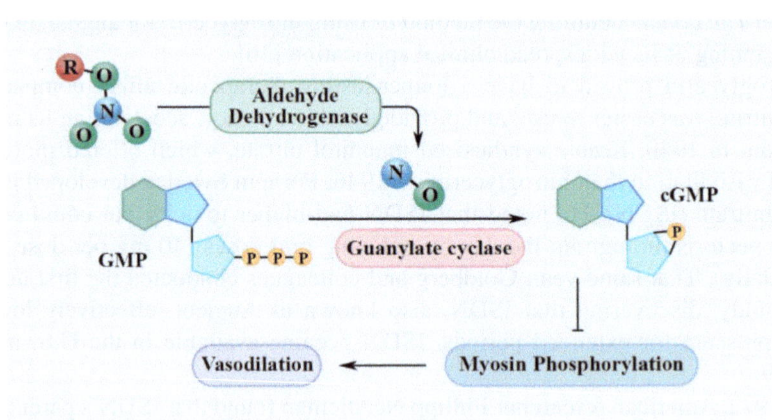

Fig. 1.10 Molecular mechanism of nitrate-induced vasodilation

Nitroglycerin is clinically administered via sublingual tablets, aerosols, intravenous solutions, immediate-release oral formulations, extended-release oral formulations, and transdermal patches. Its medical potential continues to be explored. As a nitric oxide (NO) donor, nitroglycerin was found in 2017 to inhibit postmenopausal osteoporosis without adverse effects, offering an alternative to estrogen therapy [40]. In cardiology, intermittent short-acting nitroglycerin treatments were shown in 2018 to promote the formation of collateral circulation in coronary arteries, reducing the extent of experimental myocardial infarction [41]. In 2019, it was reported that oral nitroglycerin can rapidly restore esophageal patency in cases of food-induced esophageal obstruction [42]. By 2022, emergency departments used intravenous nitroglycerin as an effective vasodilator; lower doses caused venous dilation, while higher doses led to arterial dilation [43]. Nitroglycerin remains the most commonly used drug for acute angina, typically administered as sublingual tablets or sprays [44]. However, it has limitations, including tolerance issues in which long-term exposure can significantly weaken or eliminate its therapeutic effects on angina and ischemia [45].

Ironically, Nobel suffered from heart disease late in life and was prescribed nitroglycerin. In a letter written before his death in 1896, he remarked, "Isn't it the irony of fate that I have been prescribed nitroglycerin! To avoid alarming chemists and the public, they call it Trinitrin [24]."

Nobel had 355 patented inventions in his lifetime, of which 129 were related to explosives, accumulating a wealth of $9.2 million. In his will, he stipulated that the majority of this money would be used as principal and an annual interest of $200,000 would be provided to support individuals worldwide who have made outstanding contributions in the fields of physics, chemistry, physiology or medicine, literature, and peace. The first Nobel Prize was awarded in 1901.

One hundred and two years after Nobel's death, in 1998, Robert F. Furchgott, Louis J. Ignarro, and Ferid Murad won the Nobel Prize in Physiology or Medicine for discovering NO as a signaling molecule in the cardiovascular system [45, 46]. Their work revealed how nitroglycerin drugs act through NO on the cardiovascular system, paying homage to Nobel's legacy by highlighting how his wealth, derived from nitroglycerin, contributed to scientific advancement. Despite being replaced by more powerful explosives in its original field, nitroglycerin continues to find new applications in medicine.

1.4 Application Status and Prospect of Nitrates

Wen Pan

Due to many historical reasons and various controversies, certain nitrates, including saltpeter, which can be used to make gunpowder, are listed as controlled substances, and their clinical application is dying out.

Nitrates ingested through the mouth are absorbed in the gastrointestinal tract into the blood and enter the systemic circulation. About 25% of the nitrates in the blood are taken up by the salivary glands and transported to the saliva, making the concentration of nitrate in the saliva about 10 times higher than that in the blood. Based on the physiological phenomenon that nitrate concentration in saliva is significantly higher than that in blood, the research group of Wang Songling, an academician of the Chinese Academy of Sciences, identified the parotid gland as the main organ among the four groups of salivary glands that transport nitrates from blood to saliva. The team then used the parotid gland as a model organ, to discover the presence of sialin, a nitrate transport channel, in human cell membranes. Further studies have found that nitrates play an important role in cell regeneration, cell metabolism, immune regulation, and disease prevention through nitric oxide and sialin.

Nitrates have significant medicinal value. At present, they have been added to energy drinks as a natural dietary nutrient to regulate human function. How to apply nitrates to clinical treatment and chronic disease prevention is an issue that deserves attention. The short half-life of nitrates in the human body, their low bioavailability, and the difficulty in maintaining effective blood concentration are the bottlenecks that hinder the clinical application of nitrates. Research and development of new nitrates compound preparations that are more stable, maintain longer efficacy, and are more easily absorbed by the body is a technical barrier that needs to be broken through. Safety evaluation of drugs is an essential part of drug development.

With the continuous deepening of research on new nitrate preparations in various organs and tissues, and the continuous exploration of beneficial effects, nitrates, serving as a messenger from the mouth to the whole body, are expected to offer even greater benefits to mankind.

Compliance with Ethical Standards This chapter does not contain any studies with animals performed by any of the authors.

References

1. Nanjing University of Chinese. Medicine dictionary of traditional Chinese medicine. Shanghai: Shanghai Scientific and Technical Publishers; 2014.
2. Wu P. Shennong Bencao Jing, Shennong's classic of Materia Medica. Changchun: Times Literature and Art Publishing House; 2008.
3. Mawangdui Group for Sorting Out Silk Books of Han Tombs. Mawangdui silk books of Han tombs. Beijing: Cultural Relics Publishing House; 1985.
4. Chen H. Focus and inspiration of discussions about compilation time of Wu-Shi-Er-Bing-Fang. J Chengdu Univ Tradit Chin Med. 2014;37(4):110–2.
5. He A. Differentiation of Chun Yuyi's life and deeds. Literature. 1988;2:102–13.
6. Qian S. Redord of the grand history of China. Changsha: Yue Lu Book Society; 2005.
7. Zhao K, Zhao Y. The history of distinguishing between saltpeter and saltpeter in ancient China. Res Hist Nat Sci. 1994;13(4):336–49.
8. Zhang Z, Jingui Y. Essential prescriptions from the Golden cabinet. Fuzhou: Fujian Science and Technology Publishing House; 2011.
9. Gansu Provincial Museum. Wuwei Han dynasty medical bamboo slips. Beijing: Cultural Relics Publishing House; 1975.

10. Shanghai Ancient Books Publishing House, National Library of France. Collection of Dunhuang and Western Regions documents. Shanghai: Shanghai Ancient Books Publishing House; 1998.
11. Meng N, Lv Y, Li X. Experimental study on the Chinese alchemy techniques of sulfur, saltpeter, and cinnabar. Res Hist Nat Sci. 1984;3(2):113–27.
12. Tao H. Bencaojing Jizhu. Beijing: People's Health Publishing House; 1994.
13. Chi L, Wang J, Chen T, et al. Efficacy research of niter based on ancient medical books. Chin Mod Tradit Chin Med. 2022:1673-4890.
14. Chang M. Compilation and annotations of Japanese Materia Medica. Beijing: China Medical Science and Technology Press; 2016.
15. Shang Z. Rihuazi and "Rihuazi Bencao". Jiangsu Tradit Chin Med. 1998;19(12):3–5.
16. Tang S. Zhenglei Bencao. Beijing: Huaxia Publishing House; 1993.
17. Chi L, Zhou J, Wang J, et al. Ancient application of nitrates in oral diseases—research progress of nitrates in stomatology. Liaoning J Tradit Chin Med. 2023;50(10):52–8.
18. Zhu S, Fang P. Prescriptions for universal relief: Volume 9. Beijing: People's Health Publishing House; 1982.
19. Yu T. Orthodoxy of medicine. Beijing: People's Health Publishing House; 1981.
20. Yan Y. Formulas to aid the living. Beijing: China Medical Science and Technology Press; 2012.
21. Wang H. Taiping Shenghui Fang, Taiping holy prescriptions for universal relief. Beijing: People's Health Publishing House; 1958.
22. Zhao J. Sheng Ji Zonglu, general records of Holy Universal Relief, vol. 2. Beijing: People's Health Publishing House; 1962.
23. Cheng P. First aid Guangsheng collection. Beijing: China Traditional Chinese Medicine Press; 1992.
24. Remane H, Remane Y. The history of the discovery of nitroglycerin. Pharm Unserer Zeit. 2010;39(5):340–4.
25. Marsh N, Marsh A. A short history of nitroglycerine and nitric oxide in pharmacology and physiology. Clin Exp Pharmacol Physiol. 2000;27(4):313–9.
26. Ringertz N. Alfred Nobel—his life and work. Nat Rev Mol Cell Biol. 2001;2(12):925–8.
27. Holmes LC, Dicarlo FJ. Nitroglycerin: the explosive drug. J Chem Educ. 1971;48(9):573–6.
28. Nakajima T. Nobel. Translated by Hui Sheng. Beijing: Xinhua Publishing House; 1983.
29. Liu J. Nobel: the legendary wealthy king of explosives. Beijing: Tongxin Publishing House; 2011.
30. Bergengren E. Biography of Nobel. Translated by Sun Wenfang. Hunan: Hunan People's Publishing House; 1983. p. 62–3.
31. Xiao S. The great scientist and entrepreneur: Nobel. Beijing: China Pictorial Press; 2009. p. 81–7.
32. Schück H, Sohlman R. Biography of Nobel. Translated by Min Ren. Beijing: Bibliography and Literature Publishing House; 1983.
33. Fant K. The complete biography of Nobel. Translated by Wang Kang. Beijing: World Knowledge Publishing House; 2013. p. 131–2.
34. Schwatrz M. The cause, relief and prevention of headaches arising from contact with dynamite. N Engl J Med. 1946;235:541–4.
35. Brunton TL. On the use of nitrite of amyl in angina pectoris. Lancet. 1867;90(2291):97–8.
36. Murrell W. Nitro-glycerine as a remedy for angina pectoris. Lancet. 1879;113(2890):80–1.
37. Ignarro LJ. Biological actions and properties of endothelium-derived nitric oxide formed and released from artery and vein. Circ Res. 1989;65(1):1–21.
38. Ignarro LJ, Byrns RE, Buga GM, et al. Endothelium-derived relaxing factor from pulmonary artery and vein possesses pharmacologic and chemical properties identical to those of nitric oxide radical. Circ Res. 1987;61(6):866–79.
39. Jaffrey SR, Snyder SH. Nitric oxide: a neural messenger. Annu Rev Cell Dev Biol. 1995;11:417–40.

40. Jamal SA, Hamilton CJ, Eastell R, et al. Effect of nitroglycerin ointment on bone density and strength in postmenopausal women: a randomized trial. JAMA. 2011;305(8):800–7.
41. Gatzke N, Hillmeister P, Dülsner A, et al. Nitroglycerin application and coronary arteriogenesis. PLoS One. 2018;13(8):e0201597.
42. Schimmel J, Slauson S. Swallowed nitroglycerin to treat esophageal food impaction. Ann Emerg Med. 2019;74(3):462–3.
43. Twiner MJ, Hennessy J, Wein R, et al. Nitroglycerin use in the emergency department: current perspectives. Open Access Emerg Med. 2022;14:327–33.
44. Divakaran S, Loscalzo J. The role of nitroglycerin and other nitrogen oxides in cardiovascular therapeutics. J Am Coll Cardiol. 2017;70(19):2393–410.
45. Stone R. Nobel century. At 100, Alfred Nobel's legacy retains its luster. Science. 2001;294(5541):288–91.
46. Lichtman MA. Alfred Nobel and his prizes: from dynamite to DNA. Rambam Maimonides Med J. 2017;8(3):e0035.

Nitrate: "Angels" or "Devils"

2

Zi Yang and Shaorong Li

The International Agency for Research on Cancer (IARC) has classified nitrates/nitrites as Group 2A carcinogens. The IARC does not directly classify nitrates or nitrites as carcinogens but rather lists "ingested nitrates or nitrites under conditions that result in endogenous nitrosation," owing to their conversion to nitroso compounds through epidemiological investigation and animal experiments. Although there is no clear and direct evidence that nitrates pose carcinogenic risks to humans [1], controversy persists among the public, media, and academics regarding their potential risks to humans (Fig. 2.1).

2.1 Nitrates and Gastrointestinal Tumors

Zi Yang

Nitrites react with biogenic amines, such as secondary or tertiary amines, under stringent conditions to form nitrosamines. The World Health Organization (WHO) recommends an upper limit of daily nitrite intake of 0.06–0.07 mg/kg and an upper limit of nitrate intake of 3.7 mg/kg [2]. In recent years, an increasing number of studies have found no significant correlation between nitrate intake and gastrointestinal tumors, and it has been found that dietary nitrates are mainly derived from vegetables, which contain high levels of antioxidants that inhibit the formation of nitrosamines, such as vitamin C [3]. There is no direct evidence from epidemiological studies that nitrates promote an increased risk of gastrointestinal cancers [1].

Z. Yang · S. Li (✉)
Beijing Stomatological Hospital, Capital Medical University, Beijing, China
e-mail: lishaorong@ccmu.edu.cn

S. Wang (ed.), *Inorganic Nitrate and Human Body Homeostasis*,
https://doi.org/10.1007/978-981-95-2824-0_2

Fig. 2.1 Nitrates—"angel" or "devil"

2.1.1 The Relationship Between Nitrosamines and Tumors

The molecular structure of nitrosamine is Rl(R2)N-N=O. Rl and R2 of nitrosamine can be an alkyl, cycloalkyl, aromatic ring or heterocyclic ring, which can be classified into volatile and non-volatile types; the volatile nitrosamines include N-nitrosodimethylamine (NDMA), N-nitrosodiethylamine (NDEA), N-nitrosodibutylamine (NDBA), N-nitrosopiperidine (NPIP), N-nitrosopyrrolidine (NPYR) and N-nitrosopiperidine (NPYR), and N-nitrosomorpholine (NMOR). The IARC classifies processed meats containing large amounts of nitrites and nitrosamines as Class 1 carcinogens, NDEA and NDMA as Class 2A, and NPYR, NPIP, and NDBA, as Class 2B carcinogens [4, 5].

2.1.2 Synthesis of Endogenous Nitrosamines and Influencing Factors

After ingestion of nitrates by an organism, it is absorbed into the circulatory system via the digestive tract and enriched in the salivary glands via sialin proteins. 20%–25% of nitrates are secreted in the oral cavity via saliva, where some of the salivary nitrates (5%–36%) are reduced to nitrites by bacteria [6]. Nitrites can be synthesized with H^+ under acidic conditions into the unstable HNO_2 (nitrous acid), followed by rapid decomposition into the nitrogen oxides (NO_x), such as N_2O_3 (nitrous anhydride). N_2O_3, as a nitrosating agent, can be synthesized into nitrosamine analogues by reacting with secondary amines [7].

The conversion of nitrate to nitrosamine in an organism requires extremely strict conditions: (1) the overall percentage of nitrates-to-nitrites conversion is low (1%–9%) [8]; (2) there is a certain time difference between the nitrites formed by the conversion of nitrates and the biogenic amines in the food that enter the stomach [9]; (3) the normal pH of gastric juice is 1.8–2.0, which can be increased to about 7.0 after eating, and the pH of the gastric juice in the 0.5–2 h post-meal is not suitable for nitrosamine synthesis [10], and even if the stomach reverts to a low pH after meals, the concentration of nitrites at this time does not support further nitrosamine synthesis; (4) amino acid groups are protected by peptide bonds during digestion, and the pepsin in the stomach only destroys the peptide bonds of tyrosine or phenylalanine, so that there is no excessive release of free amino acids from the stomach, and nitrosamine synthesis of the substrate is limited t11. Therefore, it is difficult for ingested nitrates to form nitrosamines in the body under normal conditions.

The process of endogenous nitrosamine synthesis is influenced by a number of complex factors. Nitrosation of secondary amines is most intense at acidic pH, so excessive gastric acid may promote endogenous nitrosamine synthesis [11]. *Helicobacter pylori* infection, which has been associated with gastrointestinal cancers, has been found to cause an increase in gastric acid, leading to a decrease in pH and a significant increase in nitrate and nitrite concentrations in the gastric juice [12]. The higher the degree of inflammation in gastritis, the higher the level of the nitrite and the lower the vitamin C within the gastric juice [13, 14], and the endogenous nitrosamine concentration synthesized may be associated with the inflammatory state. Patients with inflammatory bowel disease have increased intestinal nitroso compound synthesis, which may be associated with colonic nitric oxide synthase (NOS) activity [15]. Based on previous work, the group of academician Wang Songling synthesized a new drug, "Nitrator," by combining nitrate with vitamin C, which further reduces the possibility of nitrite formation from nitrates in the body.

After nitrates enter the body, 75%–80% of them are excreted directly through urine and sweat, and 1%–9% are converted to nitrites. At a specific pH in the stomach, they may react with concomitantly ingested biogenic amines to form endogenous nitrosamines, a process governed and influenced by a variety of factors.

2.1.3 Relationship Between Nitrites and Gastrointestinal Tumors

The relationship between dietary nitrites and gastrointestinal tumors is still controversial. The carcinogenicity of nitrites was reported as early as 1958 [16]. However, some studies have found that dietary nitrites, especially from animal sources, are positively associated with the growth of gastrointestinal tumors, but concomitant intake of vitamin C significantly reduces the risk of cancer development. In addition, some studies have also reported that no significant correlation was seen between dietary nitrites and gastrointestinal cancers (Table 2.1).

Table 2.1 Clinical studies on the correlation between nitrites and gastrointestinal tumors

Year	Study type	Region	Sample size	Results
1990 [17]	Case-control study	Italy	Gastric cancer group: 1016 Control group: 1159	The risk of gastric cancer was significantly higher with increased intake of nitrites and proteins, and lower with increased intake of ascorbic acid, B-carotene, a-tocopherol, and vegetable fats.
1999 [18]	Cohort study	Finland	9985	Dietary nitrite intake was not significantly associated with colorectal cancer incidence. Nitrosamine intake was significantly associated with an increased risk of colorectal cancer
2001 [19]	Case-control study	USA	Non-cardia gastric cancer: 352 cases; gastric cancer of the cardia: 225 squamous cell carcinoma of the oesophagus: 206; adenocarcinoma of the oesophagus: 282 cases	Increased nitrite intake was significantly associated with increased risk of non-cardia gastric cancer Vitamin supplementation was significantly associated with reduced risk of non-cardia gastric cancer.
2011 [20]	Cohort study	USA	303,156	Red meat intake was positively associated with oesophageal squamous cell carcinoma Nitrite intake was not significantly associated with gastrointestinal cancers.
2012 [21]	Case-control study	USA	Total: 494,979 squamous cell carcinoma of the oesophagus: 215; adenocarcinoma of the oesophagus: 630; adenocarcinoma of the gastric cardia: 454; non-cardia adenocarcinoma of the stomach: 501	Nitrites were not associated with esophageal or stomach cancer
2014 [21]	Cohort study	China	73,118	There was no significant correlation between overall nitrite intake and colorectal cancer risk
2021 [22]	Cohort study	USA	98,030 women	Nitrite intake from processed meats was associated with an increased risk of gastric cancer; dietary nitrites were negatively associated with gallbladder cancer, and nitrites from animal sources were negatively associated with small bowel cancer

2.1.4 Relationship Between Nitrates and Gastrointestinal Tumours

Professor Wang Songlin fed pigs with high doses of nitrate for 2 years and did not observe any significant changes in their gastrointestinal mucosa [23]. In 1995, the Subcommittee on Nitrate and Nitrite in Drinking Water of the U.S. National Science Research Council [24] concluded that there was no credible evidence linking nitrates to gastric cancer incidence and mortality. In 2010, the IARC also published that there is currently no substantial evidence that nitrates are carcinogenic to animals [1]. Several large-scale prospective cohort studies and case-control studies have found no significant correlation between dietary nitrate intake and gastrointestinal cancers, and some studies have found that nitrates derived from fruits or vegetables may reduce the risk of gastrointestinal tumors (Table 2.2).

In summary, nitrates are very stable in the body, with a small portion converted to nitrites and only small amounts of nitrosamines produced under stringent

Table 2.2 Clinical studies on the correlation between nitrates and gastrointestinal tumors

Year	Study type	Region	Sample size	Results
1990 [17]	Case-control study	Italy	High-risk areas for stomach cancer: 1016 Low-risk areas for gastric cancer (control): 1159	No significant association was seen between dietary nitrate intake and gastric cancer risk.
1997 [25] 1998 [26]	Prospective cohort study	Holland	Male: 58,279 Female: 62,573	Tere was no significant correlation between total nitrate intake, dietary nitrate, or drinking water nitrates and gastric cancer.
1999 [18]	Cohort study	Finland	9985	Nitrate intake was negatively associated with gastric cancer, but the difference was not statistically significant
2009 [27]	Case-control study	Mexico	Gastric cancer: 257 Control group: 478	Increased risk of gastric cancer was associated with high intake of nitrates and nitrites from animal sources; intake of nitrates or nitrites from fruits and vegetables reduced gastric cancer risk
2011 [28]	Case-control study	Italy	Gastric cancer: 382 Control group: 561	Gastric cancer risk was significantly reduced with increased nitrate intake
2014 [21]	Cohort study	China	73,118	Overall nitrate intake was not significantly associated with colorectal cancer risk; high nitrate intake was associated with colorectal cancer risk when vitamin C intake was low
2021 [22]	Cohort study	USA	98,030 women	There was no correlation between dietary nitrates and gastrointestinal tumors

conditions. Current research suggests that nitrosamines are significantly associated with gastrointestinal cancers, while the relationship between nitrites and gastrointestinal cancers remains somewhat controversial. Moreover, there is no direct evidence that nitrates are associated with an increased risk of gastrointestinal cancers.

2.2 Nitrates and Methemoglobinemia

Shaorong Li

Methemoglobinemia is a condition characterized by an elevated level of methemoglobin in red blood cells, leading to tissue hypoxia and resulting in a specific disease. It can be classified into congenital and acquired ones based on its etiology [29].

2.2.1 Devil?

Nitrites combine with hemoglobin to form methemoglobin, which loses its oxygen-carrying capacity, leading to hypoxia and potentially fatal asphyxiation. Studies suggest that an excess of nitrites in the body can react with hemoglobin, triggering methemoglobinemia [29] (Fig. 2.2). In the 1940s, early research conducted in the Midwest of America showed a correlation between the incidence of infantile methemoglobinemia and nitrate levels in the well water used for preparing formula milk.

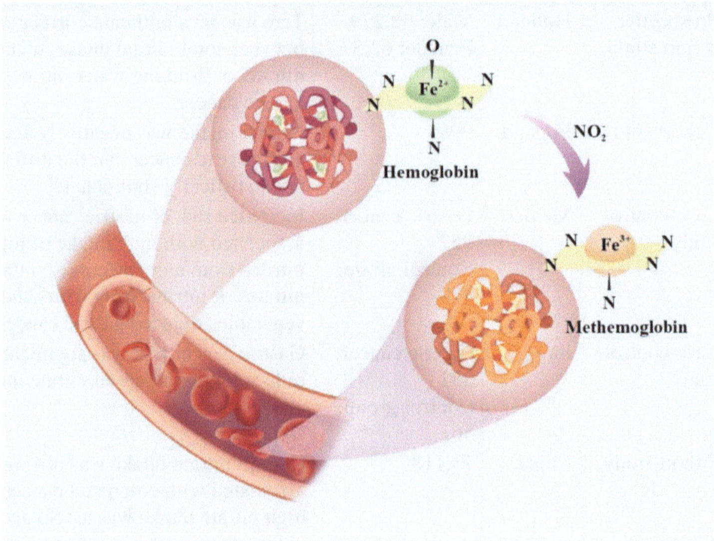

Fig. 2.2 Excessive nitrite consumption can induce methemoglobinemia

2.2.2 The Intake of Nitrates Has No Correlation with the Occurrence and Progression of Methemoglobinemia

Scholars investigating the so-called "well-water methemoglobinemia" at that time found that the wells providing water not only contained nitrates but also harbored bacteria. They noted that in every case of infants displaying cyanosis (a clinical symptom of methemoglobinemia), the wells were situated near barns or latrines. Substituting well water with municipal water led to a substantial decrease in the incidence of methemoglobinemia [30]. A reevaluation of these original studies indicated that methemoglobinemia invariably occurs when well water is contaminated with human or animal feces containing a significant number of bacteria [31]. This suggests that methemoglobinemia resulting from well water is caused by bacteria rather than by nitrates. The latest interpretation of these early studies is that the gastrointestinal irritation caused by bacteria in well water stimulates the production of nitric oxide in the intestines, which reacts with hemoglobin in the blood, converting it to methemoglobin [32].

Scholars conducted experiments on infants in the early years that closely resembled present-day nitrate toxicology tests. Researchers administered 175–700 mg of nitrate orally to the infants daily, and at that dosage, the proportion of hemoglobin converted to methemoglobin was very small. This indicated that nitrates alone do not lead to methemoglobinemia [33].

Furthermore, other scholars have reported another common cause of infantile methemoglobinemia—infectious enteritis, which can lead to an increase in endogenous nitric oxide production. This suggested that many cases of infantile methemoglobinemia once thought to be caused by nitrates in well water were actually due to gastrointestinal infections [34]. Studies also indicated that even when breastfeeding mothers consume significant amounts of nitrates (100 mg/L), it did not lead to the occurrence of infantile methemoglobinemia [35].

Considering the occurrence of methemoglobinemia, the American Public Health Association currently provided recommended upper limits for nitrate intake. However, even the authors of this report acknowledged the influence of incomprehensive data and methodological biases. For instance, in many cases, water samples for nitrate analysis were taken several months after the occurrence of infantile methemoglobinemia. Therefore, there is currently no definitive evidence suggesting a direct link between nitrates and methemoglobinemia.

Compliance with Ethical Standards This chapter is a review of previously published accounts, as such, no animal or human studies were performed.

References

1. IARC Working Group on the Evaluation of Carcinogenic Risks to Humans. IARC monographs on the evaluation of carcinogenic risks to humans. IARC Monogr Eval Carcinog Risks Hum. 2010;94:1–412.
2. Speijers G. Nitrite and potential endogenous formation of N-nitroso compounds. WHO Food Addit Ser. 2003;50:49–74.
3. Wiseman M. The second World Cancer Research Fund/American Institute for Cancer Research expert report. Food, nutrition, physical activity, and the prevention of cancer: a global perspective. Proc Nutr Soc. 2008;67(3):253–6.
4. Gushgari AJ, Halden RU. Critical review of major sources of human exposure to N-nitrosamines. Chemosphere. 2018;210:1124–36.
5. IARC Working Group on the Evaluation of Carcinogenic Risks to Humans. Red meat and processed meat. IARC Monogr Eval Carcinog Risks Hum. 2018;114:1–506.
6. Lundberg JO, Carlström M, Weitzberg E. Metabolic effects of dietary nitrate in health and disease. Cell Metab. 2018;28(1):9–22.
7. Mirvish SS. Formation of N-nitroso compounds: chemistry, kinetics, and in vivo occurrence. Toxicol Appl Pharmacol. 1975;31(3):325–51.
8. EFSA Panel on Food Additives and Nutrient Sources Added to Food (ANS), Mortensen A, Aguilar F, et al. Re-evaluation of sodium nitrate (E 251) and potassium nitrate (E 252) as food additives. EFSA J. 2017;15(6):e04787.
9. Govoni M, Jansson EA, Weitzberg E, et al. The increase in plasma nitrite after a dietary nitrate load is markedly attenuated by an antibacterial mouthwash. Nitric Oxide. 2008;19(4):333–7.
10. Cai LF, Li N, Du S, et al. Progress in hazards, synthesis and inhibition of N-Nitroso compounds in vitro and in vivo. Food Sci. 2016;37(5):271–7.
11. GM MK, Duncan CW, Leifert C, et al. Dietary nitrate in man: friend or foe? Br J Nutr. 1999;81(5):349–58.
12. Shiotani A, Iishi H, Kumamoto M, et al. Helicobacter pylori infection and increased nitrite synthesis in the stomach. Inflammation and atrophy connections. Dig Liver Dis. 2004;36(5):327–32.
13. Zhang ZW, Patchett SE, Perrett D, et al. The relation between gastric vitamin C concentrations, mucosal histology, and CagA seropositivity in the human stomach. Gut. 1998;43(3):322–6.
14. Kobayashi J. Effect of diet and gut environment on the gastrointestinal formation of N-nitroso compounds: a review. Nitric Oxide. 2018;73:66–73.
15. Pliss GV. Carcinogenic activity of dicyclohexylamine and of its nitrite salts. Vopr Onkol. 1958;4(6):659–69.
16. Buiatti E, Palli D, Decarli A, et al. A case–control study of gastric cancer and diet in Italy:II. Association with nutrients. Int J Cancer. 1990;45(5):896–901.
17. Knekt P, Järvinen R, Dich J, et al. Risk of colorectal and other gastro-intestinal cancers after exposure to nitrate, nitrite and N-nitroso compounds: a follow-up study. Int J Cancer. 1999;80(6):852–6.
18. Mayne ST, Risch HA, Dubrow R, et al. Nutrient intake and risk of subtypes of esophageal and gastric cancer. Cancer Epidemiol Biomarkers Prev. 2001;10(10):1055–62.
19. Cross AJ, Freedman ND, Ren JS, et al. Meat consumption and risk of esophageal and gastric cancer in a large prospective study. Am J Gastroenterol. 2011;106(3):432–42.
20. Dellavalle CT, Xiao Q, Yang G, et al. Dietary nitrate and nitrite intake and risk of colorectal cancer in the Shanghai women's health study. Int J Cancer. 2014;134(12):2917–26.
21. Buller ID, Patel DM, Weyer PJ, et al. Ingestion of nitrate and nitrite and risk of stomach and other digestive system cancers in the Iowa women's health study. Int J Environ Res Public Health. 2021;18(13):6822.
22. Xia DS, Qu XM, Tran SD, et al. Histological characteristics following a long-term nitrate-rich diet in miniature pigs with parotid atrophy. Int J Clin Exp Pathol. 2015;8(6):6225–34.

23. National Research Council (US) Subcommittee on Nitrate and Nitrite in Drinking Water. Nitrate and nitrite in drinking water. Washington, DC: National Academy Press; 1995.
24. Van Loon AJ, Botterweck AA, Goldbohm RA, et al. Nitrate intake and gastric cancer risk: results from The Netherlands cohort study. Cancer Lett. 1997;114(1/2):259–61.
25. Van Loon AJ, Botterweck AA, Goldbohm RA, et al. Intake of nitrate and nitrite and the risk of gastric cancer: a prospective cohort study. Br J Cancer. 1998;78(1):129–35.
26. Hernández-Ramírez RU, Galván-Portillo MV, Ward MH, et al. Dietary intake of polyphenols, nitrate and nitrite and gastric cancer risk in Mexico City. Int J Cancer. 2009;125(6):1424–30.
27. Palli D, Russo A, Decarli A. Dietary patterns, nutrient intake and gastric cancer in a high-risk area of Italy. Cancer Causes Control. 2001;12(2):163–72.
28. Umbreit J. Methemoglobin—it's not just blue: a concise review. Am J Hematol. 2007;82(2):134–44.
29. Comly HH. Cyanosis in infants caused by nitrates in well water. JAMA. 1987;257(20):2788–92.
30. Avery AA. Infantile methaemoglobinaemia: reexamining the role of drinking water nitrates. Environ Health Perspect. 1999;107(7):583–6.
31. Acheson ED. Nitrate in drinking water. London: HMSO; 1985.
32. Cornblath M, Hartmann AF. Methaemoglobinaemia in young infants. J Pediatr. 1948;33(4):421–5.
33. Hegesh E, Shiloah J. Blood nitrates and infantile methaemoglobinaemia. Clin Chim Acta. 1982;125(2):107–15.
34. Dusdieker LB, Stumbo PJ, Kross BC, et al. Does increased nitrate ingestion elevate nitrate levels in human milk? Arch Pediatr Adolesc Med. 1996;150(3):311–4.
35. Phillips WE. Naturally occurring nitrate and nitrite in foods in relation to infant methaemoglobinaemia. Food Cosmet Toxicol. 1971;9(2):219–28.

Correction of "Nitro"

3

Zi Yang

Nitrates come from a wide range of sources and are usually metabolized through dietary intake and the nitrate gastrointestinal-salivary cycle. Nitrates play an important role in maintaining homeostasis through nitric oxide and "sialin" and are essential substances that can perform a variety of physiological functions.

3.1 Sources of Nitrates

Nitrates are widely present in water, soil, air, plants, food, etc. The body obtains nitrates through two main routes: exogenous nitrate intake and endogenous nitrate production [1]. Dietary intake is the main source of nitrates [1, 2].

3.1.1 Exogenous Nitrate Intake

Exogenous nitrates are mainly ingested through food, with vegetables being the main source of nitrates, accounting for 80%–90% [2, 3]. The U.S. National Academy of Sciences [4] reported that 87% of food-related dietary nitrates come from vegetables. Nitrate-rich vegetables are generally in the cruciferae (e.g., arugula, radish, mustard), quinoa (e.g., beetroot, swiss chard, spinach), and amaranthaceae families. In addition, the Asteraceae (lettuce) and Umbelliferae (celery and parsley) families include many species with high nitrate content (Table 3.1) [3].

Other sources of nitrate intake include drinking water (15%) and other foods (5%) [5], and the nitrate content is subject to strict standards. Nitrates are naturally present in water supplies; however, in most countries, the concentration of nitrates in water is usually well below the permitted concentration ($\leq$50 mg/L) [6]. Nitrates

Z. Yang (✉)
Beijing Stomatological Hospital, Capital Medical University, Beijing, China

S. Wang (ed.), *Inorganic Nitrate and Human Body Homeostasis*,
https://doi.org/10.1007/978-981-95-2824-0_3

Table 3.1 Nitrate concentrations in vegetables

Nitrate concentration	Species
Very low (<200 mg/kg fresh weight)	Artichokes, asparagus, broad beans, aubergines, garlic, onions, green beans, mushrooms, peas, peppers, potatoes, courgettes, sweet potatoes, tomatoes, watermelon
Low (200–500 mg/kg fresh weight)	Broccoli, carrots, cauliflower, cucumber, pumpkin, "Puntarelle" chicory
Medium (500–1000 mg/kg fresh weight)	Cabbage, dill, turnip, savoy cabbage
High (1000–2500 mg/kg fresh weight)	Celeriac, Chinese cabbage, leaf chicory, fennel, kohlrabi, leek, parsley
Very high (>2500 mg/kg fresh weight)	Celery, chervil, cress, lettuce. Red beetroot, spinach, swiss chard

Table 3.2 Nitrate content of meat products

	Nitrate content/mg/kg	
Name of meat product	Average value	Range
Salami	94	Not detected–450
Hot dog (loanword)	64	8–81
Tinned pickled meat	63	Not detected–840
Sausages	58	15–240
Hams	55	Not detected–1400
Bacon (loanword)	42	Not detected–310
Luncheon meat	32	<10–70
Corned beef	14	4–36
Minced beef	12	Not detected–24

and nitrites (e.g., potassium nitrite/sodium nitrate) have been used for many years as food additives in cured meats (Table 3.2) [7]. Therefore, nitrates from the water supply and processed meat products contribute little to total nitrate intake compared to that from vegetable sources.

3.1.2 Endogenous Nitrate Production

In 1978, researchers published a paper in *Science*. They gave six volunteers either a protein-free diet or a 0.8 g/kg/day protein diet, calculated the average total amount of nitrates ingested and excreted by the volunteers, and found that, regardless of diet, the total amount of nitrates excreted in the urine was significantly higher than the total amount of nitrates ingested. Their findings demonstrated that endogenous synthesis of nitrate occurs in the body and that the main site of its synthesis was found to be the small intestine [8]. Current research suggests that endogenous nitrate synthesis in mammals accounts for 10%–20% of the total [2, 3], and that its production is mainly from the oxidation of nitrites and NO. High concentrations of salivary nitrates are partially reduced to nitrites and NO by oral bacteria. Nitrates and nitrites can be absorbed into the bloodstream via the intestinal mucosa, and nitrites can be converted back to nitrates. In addition, during the production of NO from L-arginine, the nonessential amino acid L-arginine is converted to L-citrulline and NO in the presence of NO synthase, and the NO formed is involved in several reactions, where nitrates and nitrites are formed as by-products of the reaction [9, 10].

3.2 Nitrate Metabolism

Nitrates are absorbed after ingestion and metabolized through the nitrate gastrointestinal-salivary cycle to perform important physiological functions (Fig. 3.1).

3.2.1 Nitrate Uptake

Dietary nitrates are absorbed into the circulatory system via the digestive tract, where the bioavailability of nitrate in cooked spinach, raw lettuce and cooked beetroot is approximately 100%, and plasma nitrate concentrations peak 1 h after ingestion of the meal [11], with some studies reporting nitrate peaking times of 1.5–1.8 h

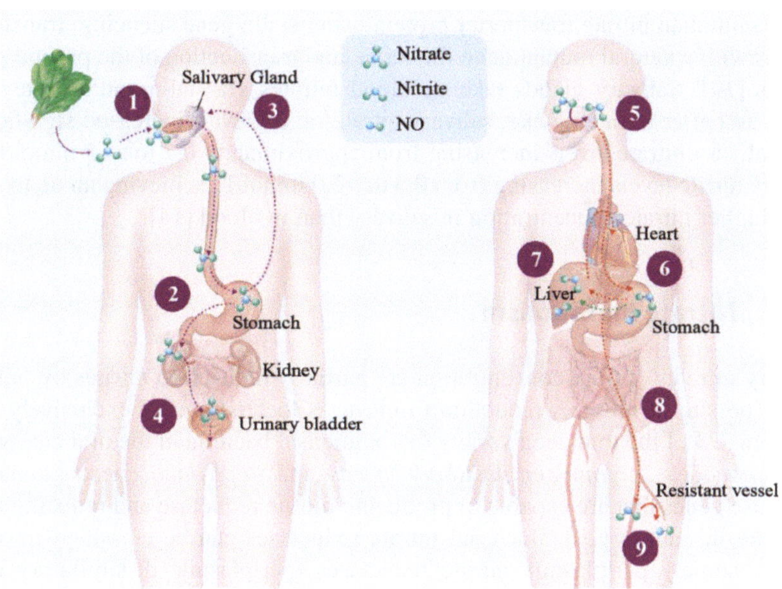

Fig. 3.1 Metabolism of nitrate. (1) Nitrate is ingested from dietary sources; (2) Nitrate is absorbed by the stomach and small intestine; (3) Nitrate is concentrated in saliva; (4) Nitrate is excreted by the kidneys; (5) Nitrate is reduced to nitrite by bacteria in the oral cavity; (6) Nitrite is reduced to NO; (7) Nitrite and NO diffuse into the portal system, and NO is oxidized to nitrite; (8) Nitrite is transported in the arterial circulation; and (9) Nitrite is reduced in the resistance vessels to NO, dilating blood vessels to lower blood pressure.

[12]. Nitrites have a bioavailability of 95%–98% after ingestion [13], and plasma nitrites increase after 15 min of nitrate ingestion and continues to increase for the next 3 h [14]. Only <1% of ingested nitrates reach the colon and are excreted in the faeces [15]. Moreover, there is no significant first-pass effect of dietary intake of nitrates and nitrites.

3.2.2 Nitrate Distribution

The volume of distribution of nitrates and nitrites is smaller than that of body water and larger than that of blood, indicating that nitrates are distributed throughout the body [16, 17]. Nitrites are rapidly absorbed into most tissues [18]. Nitrates, on the other hand, are significantly enriched in the salivary glands, with 20%–28% of them being secreted in the oral cavity via saliva. A study has clarified that the parotid gland is the main organ of nitrate transport in the organism through a pig parotid atrophy model, and cell membrane nitrate channels have been discovered based on salivary gland organs. Nitrates induce a specific cell membrane current in salivary gland cells, and this current increases exponentially in acidic environment, suggesting that nitrates enter the cell through the cell membrane and protons synergistically. Sialin was identified as a candidate protein for the cell membrane nitrate channel through genetic testing and bioinformatics analysis. It was also confirmed as a mammalian nitrate transporter protein using sialin gene silencing, transfection of cells with a natural mutant gene for sialin, and transduction of the parotid glands of pigs [19]. Salivary glands take up blood nitrates via sialin and secrete it into saliva, and after nitrate uptake, salivary nitrate/nitrite levels increased significantly, with salivary nitrate levels increasing from approximately 0.7 to >15 mmol/L, and salivary nitrite levels increasing from 0.3 to >2.0 mmol/L, achieving an up to 10–20 times higher nitrate concentration in salivary than in blood [14].

3.2.3 Nitrate Metabolism

Salivary nitrates of high concentration are partially reduced to nitrites by oral bacteria. The site of nitrate's reduction to nitrite is located almost exclusively in the posterior 1/3 of the tongue; a variety of commensal bacteria in the oral cavity, such as *Veronococcus*, *Actinobacillus*, *Rhodococcus*, and *Staphylococcus*, contain nitrate reductase genes that are capable of producing nitrate reductase and reducing nitrate to nitrite in culture [20]. Bacterial nitrate reductases can be classified into three groups, namely periplasmic nitrate reductases (periplasmic dissimilatory nitrate reductases, Nap), membrane-bound nitrate reductases (membrane-bound respiratory nitrate reductases, Nar), and cytoplasmic assimilatory nitrate reductases (Nas) [20, 21].

While dietary nitrates can also be converted to nitrites by nitrate reductase on the surface of the tongue before they are swallowed, the vast majority of nitrites are formed subsequently through the gastrointestinal-salivary cycle of nitrate, where nitrates accumulate in the salivary glands and are secreted into saliva [19]. Nitrites, along with the remaining nitrates, are swallowed into the stomach and further converted to NO and other biologically active nitrogen oxides in an acidic environment. When endogenous production of NO via NOS is limited, nitrites can be reduced to NO or stored in blood and tissues under certain physiological conditions (e.g., hypoxia) [22], and the residual nitrates and nitrites can subsequently be absorbed through the small intestine.

3.2.4 Nitrate Excretion

Most of the ingested nitrates (65%–75%) are excreted by the kidneys [10]. Nitrates have a half-life of 5–8 h while nitrites have a half-life of 1–5 min in in vitro experiments [23] and 20–45 min in vivo [13, 17]. The above suggests that nitrates are highly stable in the circulation and that nitrites tend to oxidize to nitrates or reduce to NO under oxygenated conditions.

3.3 Nitrates—Essential Substances for the Human Body

Nitrates play an important role in the maintenance of organismal homeostasis through NO and sialin. Nitrate ingestion activates the NO_3^-–NO_2^-–NO pathway, which is a parallel pathway and a powerful complement to the classical NO pathway. Meanwhile, nitrates can also play a role in intracellular transport via sialin protein. They can perform a variety of functions in the organism, including improving vascular endothelial function, regulating vascular tone, exerting antioxidant effects, inhibiting the release of inflammatory factors, regulating glycolipid metabolism, and regulating intestinal flora, thus improving muscle motility, protecting the digestive system and cardiovascular system, and alleviating systemic disorders, such as stress ulcers and pulmonary arterial hypertension (Fig. 3.2) [19].

3.3.1 Nitrates Prevent Cardiovascular Disease

Chronic dietary nitrate deficiency in mice can lead to endothelial dysfunction and cardiovascular disease, among others [24]. Using beetroot juice as a source of nitrates in healthy volunteers, a reduction in blood pressure of about 10 mmHg was found in healthy volunteers, which coincided with peak plasma nitrite levels [11]. A randomized controlled crossover trial involving 15 hypertensive patients (treated) found that antibacterial mouthwash usage for more than 3 days resulted in a 2.3 mmHg increase in systolic blood pressure but had no significant effect on diastolic blood pressure [25]. This suggests the importance of the nitrate gastrointestinal-salivary cycle and highlights the key role of oral nitrate-reducing bacterial bioactivation, especially in the context of dysfunction of the L-arginine-NOS pathway, where nitrates and nitrites are regarded as biologically active dietary constituents and indispensable nutrients for the maintenance of NO homeostasis [19].

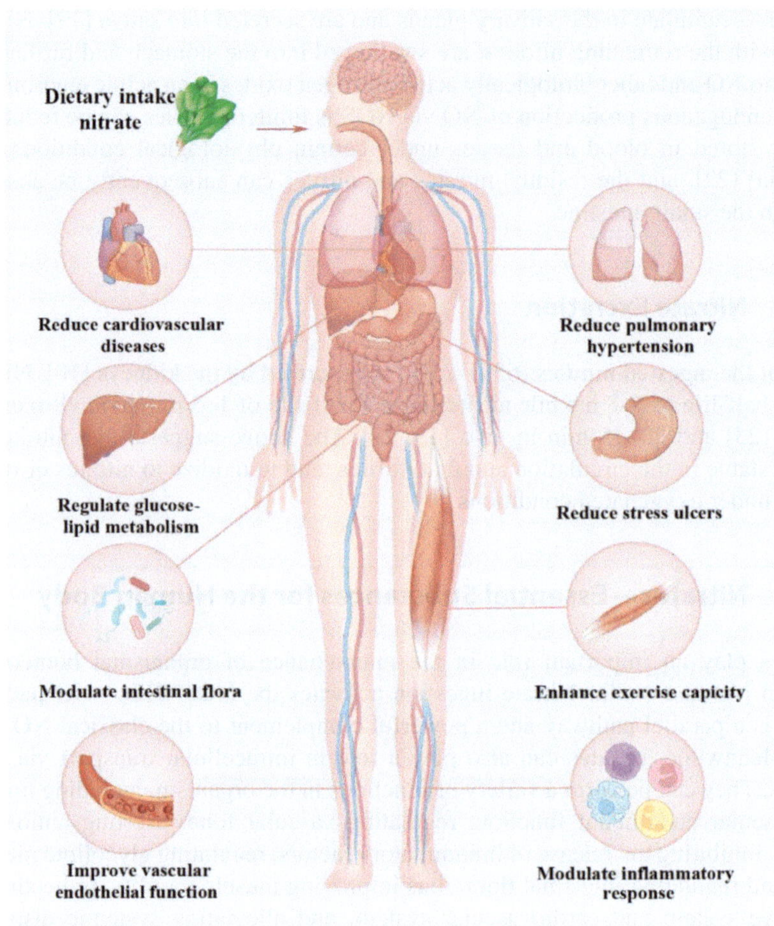

Fig. 3.2 Physiological functions of nitrate

3.3.2 Nitrates Improve Salivary Gland Function

Our group found that elevated nitrate concentration in the blood is usually accompanied by increased sialin expression in important organs, and that nitrates can increase sialin expression in alveolar cells, which further promotes the entry of nitrates into the cells. Nitrate intake increased the expression of salivary gland water channel protein AQP5, increased salivary secretion, and effectively reduced the fibrotic area and cell shrinkage of salivary gland tissue. This process was mediated by a reciprocal nitrate-sialin feedback loop, which promoted proliferation of alveolar and ductal cells and reduced apoptosis by upregulating the EGFR-AKT-MAPK signaling pathway [26]. In addition, our group also found that the nitrate-sialin system can preserve salivary gland cells from radiation-induced damage by regulating intracellular autophagy in salivary gland cells.

3.3.3 Nitrates Improve Exercise Capacity

The role of NO in regulating mitochondrial function and energy expenditure is of great interest. Dietary nitrate supplementation in healthy young volunteers reduces oxygen consumption and improves muscle efficiency during a variety of submaximal exercises (exercise loads of 70%–85% of the maximum exercise load) without increasing anaerobic metabolism [27]. Cyclists given a beetroot juice drink for three consecutive days exhibited a one-fold increase in plasma nitrite concentration and a reduction of approximately 5% in oxygen consumption during exercise [28]. Nitrate-rich diets reduce the initial muscle discharge rate at rest as well as the mean and maximum discharge rates during fatiguing exercise and may enhance exercise function by improving muscle and nerve function and oxygen consumption [29]. A previous study by our group found that nitrates can upregulate sialin expression in MSCs, thereby improving mitochondrial function and reducing their senescence levels.

3.3.4 Nitrates Protect the Digestive System

Gastrointestinal protection is the earliest evidence that nitrates are beneficial to the human body. Salivary glands actively secrete nitrates to protect the gastrointestinal tract when the body is under stress [30]. Nitrate supplementation in rats resulted in a 20% increase in the thickness of the firmly adherent mucus layer, and the protective effect of nitrate supplementation on the gastrointestinal tract was blocked by the use of antimicrobial oral sprays [31]. Our group found that nitrates can combat inflammatory bowel disease and balance intestinal flora by elevating NO levels [32]. Meanwhile, nitrate can directly regulate the macrophage M1/M2 ratio through the sialin protein, thus preventing NAFLD.

3.3.5 Nitrates Regulate Metabolism

Compared to a conventional diet, a low-nitrite/nitrate diet in mice significantly triggers visceral obesity, dyslipidemia, and glucose intolerance at 3 months; it also induces body weight gain, hypertension, insulin resistance, and impaired impaired endothelium-dependent relaxations to acetylcholine at 18 months, and significantly results in death due to cardiovascular disease at 22 months [33]. NO_3^-–NO_2^-–NO pathway targets include oxidative stress and adipose tissue browning, with activation of typical metabolic regulatory pathways similar to those affected by several currently used glucose-lowering drugs [24].

Overall, nitrates play an important role in the organism, whose deficiency causes changes in cardiovascular, gastrointestinal, and metabolic homeostasis. Nitrates can maintain homeostasis by activating the nitrate-nitrite-NO pathway and regulating the sialin protein, and many of the physiological functions of nitrates still need further investigations.

Compliance with Ethical Standards This chapter is a review of previously published accounts, as such, no animal or human studies were performed.

References

1. Weitzberg E, Lundberg JO. Novel aspects of dietary nitrate and human health. Annu Rev Nutr. 2013;33:129–59.
2. Hord NG, Tang YP, Bryan NS. Food sources of nitrates and nitrites: the physiologic context for potential health benefits. Am J Clin Nutr. 2009;90(1):1–10.
3. Pietro S. Nitrate in vegetables: toxicity, content, intake and EC regulation. J Sci Food Agric. 2006;86(1):10–7.
4. National Academy of Sciences. The health effects of nitrate, nitrite and N-nitroso compounds. Washington, DC: The National Academies Press; 1981.
5. Sindelar JJ, Milkowski AL. Human safety controversies surrounding nitrate and nitrite in the diet. Nitric Oxide. 2012;26(4):259–66.
6. Ward MH, Dekok TM, Levallois P, et al. Workgroup report: drinking-water nitrate and health-recent findings and research needs. Environ Health Perspect. 2005;113(11):1607–14.
7. Jackson J, Patterson AJ, Macdonald-Wicks L, et al. The role of inorganic nitrate and nitrite in CVD. Nutr Res Rev. 2017;30(2):247–64.
8. Tannenbaum SR, Fett D, Young VR, et al. Nitrite and nitrate are formed by endogenous synthesis in the human intestine. Science. 1978;200(4349):1487–9.
9. Leaf CD, Wishnok JS, Tannenbaum SR. L-arginine is a precursor for nitrate biosynthesis in humans. Biochem Biophys Res Commun. 1989;163(2):1032–7.
10. Merino L, Örnemark U, Toldrá F. Analysis of nitrite and nitrate in foods: overview of chemical, regulatory and analytical aspects. Adv Food Nutr Res. 2017;81:65–107.
11. Webb AJ, Patel N, Loukogeorgakis S, et al. Acute blood pressure lowering, vasoprotective, and antiplatelet properties of dietary nitrate via bioconversion to nitrite. Hypertension. 2008;51(3):784–90.
12. Van Velzen AG, Sips AJ, Schothorst RC, et al. The oral bioavailability of nitrate from nitrate-rich vegetables in humans. Toxicol Lett. 2008;181(3):177–81.
13. Hunault CC, Van Velzen AG, Sips AJ, et al. Bioavailability of sodium nitrite from an aqueous solution in healthy adults. Toxicol Lett. 2009;190(1):48–53.
14. Govoni M, Jansson EA, Weitzberg E, et al. The increase in plasma nitrite after a dietary nitrate load is markedly attenuated by an antibacterial mouthwash. Nitric Oxide. 2008;19(4):333–7.
15. Bartholomew B, Hill MJ. The pharmacology of dietary nitrate and the origin of urinary nitrate. Food Chem Toxicol. 1984;22(10):789–95.
16. EFSA Panel on Food Additives and Nutrient Sources Added to Food (ANS), Mortensen A, Aguilar F, et al. Re-evaluation of sodium nitrate (E 251) and potassium nitrate (E 252) as food additives. EFSA J. 2017;15(6):e04787.
17. Dejam A, Hunter CJ, Tremonti C, et al. Nitrite infusion in humans and nonhuman primates: endocrine effects, pharmacokinetics, and tolerance formation. Circulation. 2007;116(16):1821–31.
18. Bryan NS, Fernandez BO, Bauer SM, et al. Nitrite is a signaling molecule and regulator of gene expression in mammalian tissues. Nat Chem Biol. 2005;1(5):290–7.
19. Qing LZ, Jin LY, Qu XM, et al. Nitrate: a pioneer from the mouth to the systemic health and disease. Chin J Stomatol. 2020;55(7):433–8.
20. Koch CD, Gladwin MT, Freeman BA, et al. Enterosalivary nitrate metabolism and the microbiome: intersection of microbial metabolism, nitric oxide and diet in cardiac and pulmonary vascular health. Free Radic Bio Med. 2017;105:48–67.
21. Koopman JE, Buijs MJ, Brandt BW, et al. Nitrate and the origin of saliva influence composition and short chain fatty acid production of oral microcosms. Microb Ecol. 2016;72(2):479–92.

22. Burleigh MC, Liddle L, Monaghan C, et al. Salivary nitrite production is elevated in individuals with a higher abundance of oral nitrate-reducing bacteria. Free Radic Biol Med. 2018;120:80–8.
23. Lundberg JO, Govoni M. Inorganic nitrate is a possible source for systemic generation of nitric oxide. Free Radic Biol Med. 2004;37(3):395–400.
24. Lundberg JO, Carlström M, Weitzberg E. Metabolic effects of dietary nitrate in health and disease. Cell Metab. 2018;28(1):9–22.
25. Bondonno CP, Liu AH, Croft KD, et al. Antibacterial mouthwash blunts oral nitrate reduction and increases blood pressure in treated hypertensive men and women. Am J Hypertens. 2015;28(5):572–5.
26. Feng XY, Wu ZF, Xu JJ, et al. Dietary nitrate supplementation prevents radiotherapy-induced xerostomia. Elife. 2021;10:e70710.
27. Larsen FJ, Weitzberg E, Lundberg JO, et al. Effects of dietary nitrate on oxygen cost during exercise. Acta Physiol. 2007;191(1):59–66.
28. Olsson H, Al-Saadi J, Oehler D, et al. Physiological effects of beetroot in athletes and patients. Cureus. 2019;11(12):e6355.
29. Flanagan SD, Looney DP, Miller MJ, et al. The effects of nitrate-rich supplementation on neuromuscular efficiency during heavy resistance exercise. J Am Coll Nutr. 2016;35(2):100–7.
30. Jin LY, Qin LZ, Xia DS, et al. Active secretion and protective effect of salivary nitrate against stress in human volunteers and rats. Free Radic Biol Med. 2013;57:61–7.
31. Petersson J, Carlström M, Schreiber O, et al. Gastroprotective and blood pressure lowering effects of dietary nitrate are abolished by an antiseptic mouthwash. Free Radic Biol Med. 2009;46(8):1068–75.
32. Hu L, Jin LY, Xia DS, et al. Nitrate ameliorates dextran sodium sulfate-induced colitis by regulating the homeostasis of the intestinal microbiota. Free Radic Biol Med. 2020;152:609–21.
33. Kina-Tanada M, Sakanashi M, Tanimoto A, et al. Long-term dietary nitrite and nitrate deficiency causes the metabolic syndrome, endothelial dysfunction and cardiovascular death in mice. Diabetologia. 2017;60(6):1138–51.

Part II

Searching for Nitrate Along Oral Saliva

Nitrates are the trailblazers which enter from the oral cavity and move throughout the whole body, with important physiological functions and potential roles in disease prevention, and are expected to become the "probiotics" for the health of the whole body. The oral cavity and salivary glands play an important role in nitrate metabolism. So, how did researchers discover that nitrates play such an important role in human health and disease prevention?

It All Starts at One End

4

Meizhi Ning

Nitrate metabolism in the human body is primarily a multiorgan cycle, in which salivary glands are considered to be important organs. In normal people, salivary nitrate level is much higher than blood nitrate level, suggesting that the salivary glands do not simply excrete nitrates, but actively take up nitrates from the blood. So, why do the salivary glands actively take up nitrates and secrete them into the saliva? Of the four groups of salivary glands in the mouth, which one actively absorbs nitrates from blood and then secretes nitrates into the saliva? How are nitrates distributed in saliva, blood, and urine? How will the levels of nitrates change in saliva, blood, and urine when salivary gland function is impaired?

4.1 Experiment on Parotid Atrophy in Miniature Pigs

4.1.1 Experimental Methods

The function and anatomical structure of the parotid glands in miniature pigs are the closest to those in humans [1]; therefore, the miniature pig parotid atrophy model is suitable for investigating the role of salivary glands in nitrate metabolism. The parotid ducts of miniature pigs were injected with 4 mL of 1% methyl viologen into both sides of the parotid ducts, and the mouths of the ducts were sutured to chemically destroy the parotid glands and make them atrophy to establish a model of parotid gland destruction and atrophy in miniature pigs. Following the injection of 1% methyl viologen into the parotid glands of normal miniature pigs, the main pathological manifestations included acute glandular necrosis, glandular atrophy, and fibrous tissue repair [2–5]. Nitrate loading experiments were carried out

M. Ning (✉)
Department of Stomatology, Beijing Boai Hospital, China Rehabilitation Research Center, Beijing, China

S. Wang (ed.), *Inorganic Nitrate and Human Body Homeostasis*,
https://doi.org/10.1007/978-981-95-2824-0_4

3 months after bilateral parotid glands were atrophied in miniature pigs, and mixed saliva, serum, and urine were collected from normal and atrophy groups at different times before and after nitrate loading.

Prior to nitrate loading, both nitrate and nitrite levels were higher in the mixed saliva of normal miniature pigs than in the atrophy group at various times. Nitrate secretion was blocked in the atrophy group by destroying the parotid glands bilaterally prior to the experiment; therefore, the nitrate concentration in the mixed saliva of the atrophy group was reduced. There was no significant difference in serum and urine nitrate concentrations between the two groups.

4.1.2 Experimental Results

With nitrate loading, the mixed salivary nitrate concentration of both the normal and atrophy groups increased significantly, and the mixed salivary nitrate concentration of the normal group was significantly higher than that of the atrophy group, which peaked at 60 min after loading administration and then gradually declined, with the atrophy group declining faster than the normal group. The concentration values of the normal group were about twice as high as those of the atrophy group at all time intervals [6]. Serum nitrate concentration increased in both groups after loading, and the increase in the atrophy group was significantly greater than that in the normal group. The peak value of serum nitrate concentration in the atrophy group was 38.9 ± 1.19 ng/μL, which was much greater than that in the normal group (20.68 ± 1.20 ng/μL). The peak value appeared at 40 min after loading, which was before the peak value of nitrate concentration in saliva, and then the concentration in serum gradually decreased. The concentration gradually decreased and basically returned to the original level after 120 min. [6] The salivary nitrate concentration in the normal group was higher than the serum nitrate concentration before and after loading, indicating that nitrates were transported from the serum into the parotid gland against the concentration difference, confirming that the parotid gland actively took up nitrates. The maximum peak of serum nitrate concentration in the atrophy group after loading was greater than that of mixed salivary nitrate concentration in the same group, which may be due to the fact that when nitrate loading was increased, nitrate secretion from the parotid gland was blocked and salivary nitrate secretion was reduced, while the body failed to regulate serum nitrates, which inevitably led to an increase in serum nitrate concentration.

There was no significant difference in urinary nitrate concentration between the two groups before nitrate loading. When nitrate loading was applied, urinary nitrate concentration increased significantly in both groups, reaching a peak at about the 180th minute after loading, with the peak in the atrophy group being almost twice that of the normal group [6]. Urinary nitrate concentration increased in both groups after nitrate loading, and more nitrates were excreted in the urine in the atrophy group than in the normal group due to impaired active nitrate uptake by the parotid gland in the atrophy group.

Mixed salivary nitrite concentration was significantly higher in the normal group than in the atrophy group. There was no change in mixed salivary nitrite concentration in the normal group after nitrate loading, indicating that the ability of oral flora to convert nitrates to nitrites in the oral cavity is limited and easily saturated by high concentrations of nitrates. The nitrite concentration in the atrophy group gradually increased after nitrate loading and reached the level of the normal group at 120 min [6].

4.1.3 Experimental Significance

In the miniature pig parotid atrophy experiment, both nitrate and nitrite concentrations in saliva were reduced after parotid atrophy in miniature pigs, and both nitrate and nitrite concentrations in saliva increased after nitrate loading, suggesting that the parotid gland is an important organ for nitrate secretion. The entry of nitrates from serum into the parotid gland was opposite to the concentration difference, confirming that the parotid gland is active in the uptake of nitrates. The important role of the parotid gland in nitrate metabolism was demonstrated by experiments on parotid gland atrophy in miniature pigs.

4.2 Experiment on the Antimicrobial Effects of Nitrates

Given that oral saliva nitrate concentration is much higher than blood nitrate concentration, how do salivary nitrates affect the oral microenvironment? The oral microenvironment is a complex ecosystem containing a wide variety of bacteria. Do nitrates and nitrites have antimicrobial effects on oral pathogenic bacteria?

4.2.1 Experimental Methods

Bacterial suspensions of various common standard oral pathogens in the logarithmic growth phase were obtained by bacterial enrichment culture of the strains in vitro, and their bacterial concentration was adjusted to 109 CFU/mL. After that, different concentrations of nitrates, nitrites and bacterial suspensions were added to a 96-well plate and incubated by adjusting the pH to different states such as 7.0, 6.0, 5.0, 4.5, 4.0, and 3.0. The absorbance values (OD) after incubation were determined, and the incubated bacterial suspensions were further transferred to solid medium for incubation to determine the minimum inhibitory concentration (MIC) and minimum bactericidal concentration (MBC) for each bacterium.

4.2.2 Experimental Results

Acidified nitrites have significant bacteriostatic and bactericidal effects at normal physiological oral salivary concentrations (2–40 mmol/L). The bacteriostatic and bactericidal effects of nitrates and nitrites on six major oral pathogens (*Streptococcus mutans*, *Porphyromonas gingivalis*, *Fusobacterium nucleatum*, *Capnocytophaga gingivalis*, *Lactobacillus acidophilus*, and *Candida albicans*) in vitro were seen to vary under different acidic conditions. Nitrates had no effect on the major oral pathogens grown in vitro. *Lactobacillus acidophilus* and *Candida albicans* were highly tolerant of acidic environments and were able to grow at pH 3.0 to 7.0, whereas the remaining four bacteria (*Streptococcus mutans*, *Porphyromonas gingivalis*, *Fusobacterium nucleatum*, and *Capnocytophaga gingivalis*) could not continue to grow at pH below 4.0.

The bacteriostatic or bactericidal effect of acidified nitrites is much greater than that of unacidified nitrites. Different concentrations of nitrites at different pH values (4.0–7.0) have various degrees of inhibition on bacterial growth, and with the decline of pH value in the oral microenvironment, nitrites may have a stronger inhibitory effect (Fig. 4.1). However, they have their own specificity for each bacterial species. The six pathogens' susceptibilities to acidified nitrite can be ranked as follows: *Streptococcus mutans* > *Porphyromonas gingivalis* = *Fusobacterium nucleatum* > *Capnocytophaga gingivalis* > *Lactobacillus acidophilus* > *Candida albicans*. *Streptococcus mutans* was sensitive to acidified nitrites, which could completely kill the bacteria at pH 7.0, 6.0, 5.0, and 4.5. The minimum bactericidal

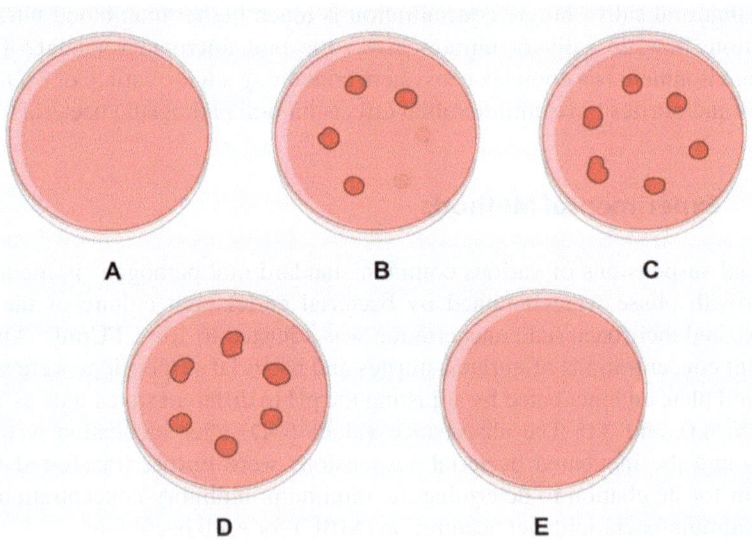

Fig. 4.1 Effects of different pH values and different concentrations of nitrates and nitrites on the growth of *Porphyromonas gingivalis*

concentration (MBC) of nitrites to inhibit *Streptococcus mutans* decreased with decreasing pH value. *Lactobacillus acidophilus* and *Candida albicans* differed in their responses to acidified nitrite compared with the other four microorganisms. In the absence of nitrite, they were significantly resistant to acid, but the addition of nitrite could inhibit their growth [7].

4.2.3 Experimental Significance

This study demonstrates that acidified nitrites have a significant effect on the growth and survival of six major oral pathogens cultured in vitro. These pathogens are strongly associated with caries, periodontitis, and halitosis; thus, it is possible that nitrate is an oral health-promoting compound, and consuming a nitrate-rich diet to increase salivary nitrite levels could be a promising therapeutic strategy to reduce caries, periodontitis, and halitosis.

4.3 Alterations of Nitrates and Nitrites in Salivary Gland Diseases

The parotid gland is the main organ involved in nitrate transport in the body, and any disease in which the parotid gland functions abnormally will result in abnormal nitrate and nitrite metabolism. Sjögren's syndrome (SS) is an autoimmune disease which primarily affects the exocrine glands, causing destruction of the salivary glands and infiltration and destruction by chronic inflammatory cells in the salivary glands, leading to a decrease in salivary secretion and affecting nitrate and nitrite metabolism.

Detection of nitrate and nitrite levels in four body fluids (parotid saliva, mixed saliva, blood, and urine) in healthy volunteers, SS patients, and patients with sialosis revealed that only nitrates but no nitrites were present in parotid saliva, suggesting that nitrites in mixed saliva are not secreted directly by the parotid gland. The highest mean concentration of nitrates in healthy volunteers was detected in parotid saliva (172 mg/L), followed by urine (160 mg/L), mixed saliva (97 mg/L), and serum (33 mg/L). The mixed saliva flow rate was lower in SS patients compared with patients with sialosis and healthy volunteers (Fig. 4.2a), and parotid saliva flow rate was much lower in SS patients (0.08 mL/min) than patients with sialosis (0.26 mL/min) and healthy volunteers (0.35 mL/min) (Fig. 4.2b). Nitrate concentration, nitrite concentration and the total amount of nitrates and nitrites in mixed saliva significantly decreased in SS patients (Fig. 4.2c, e). Nitrate concentration in parotid saliva also significantly decreased in SS patients (Fig. 4.2d). There was no difference in nitrate concentration in the mixed saliva between patients with sialosis and healthy volunteers, whereas the nitrite concentration in the mixed saliva of patients with sialosis was higher than that of both SS patients and healthy volunteers (Fig. 4.2e). This may be due to the lower flow rate of oral saliva, which promotes the

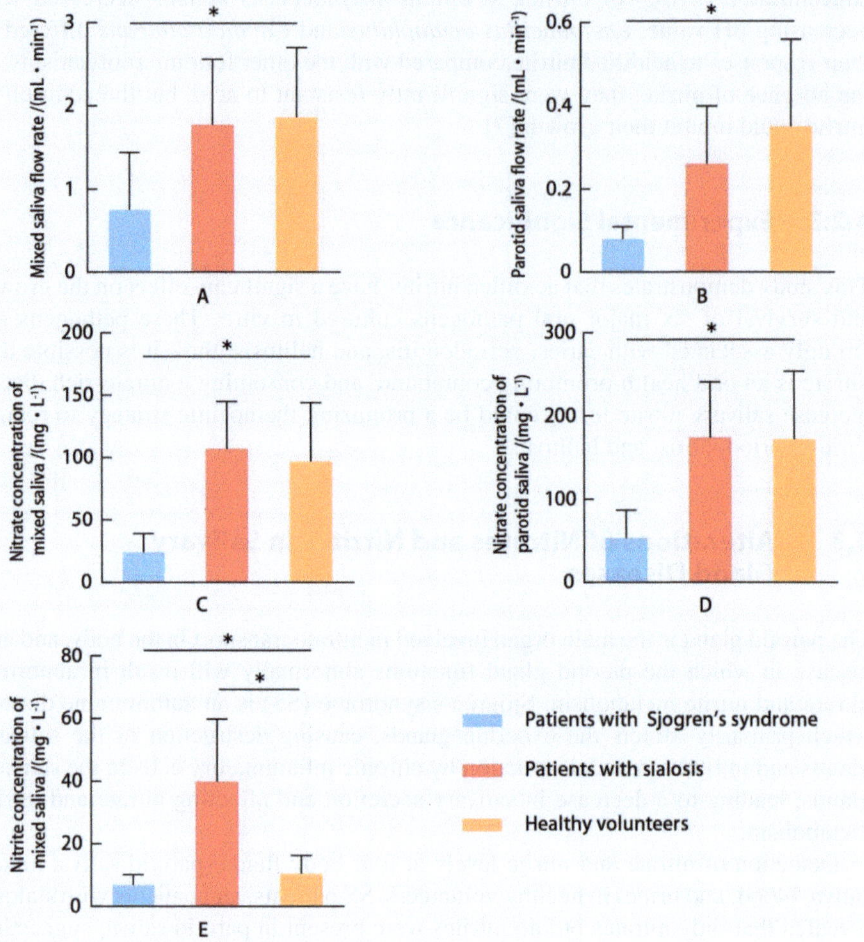

Fig. 4.2 Flow rates and nitrate and nitrite concentrations in mixed saliva and parotid saliva

growth of nitrate reductase symbiotic bacteria in the oral cavity and increases the conversion of nitrates to nitrites [8].

Nitrate concentration in parotid saliva was much higher than nitrate concentration in serum in SS patients, patients with sialosis, and healthy volunteers. Nitrate concentration in parotid saliva of patients with sialosis and healthy volunteers was six times higher than nitrate concentration in serum. Nitrate concentration in parotid saliva of SS patients was approximately two times higher than nitrate concentration in serum. The decrease of nitrate concentration in parotid saliva of SS patients may be due to salivary gland destruction by lymphocytic infiltration. Nitrate concentration in urine was comparable to that in parotid saliva in patients with sialosis and in

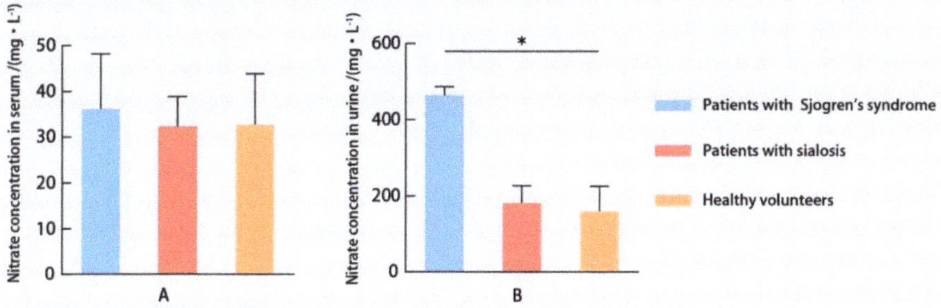

Fig. 4.3 Nitrate concentration in serum and urine

healthy volunteers, and significantly increased in SS patients, being approximately nine times the nitrate concentration in parotid saliva. Nitrate (Fig. 4.3a) and nitrite concentrations in serum remained relatively stable in all tested groups [8]. Nitrate concentration in saliva significantly decreased in SS patients, whereas nitrate concentration in serum remained stable, and nitrate concentration in urine also significantly increased (Fig. 4.3b), suggesting that nitrates are excreted from the body through urine.

Compliance with Ethical Standards All applicable international, national, and/or institutional guidelines for the care and use of animals were followed.

All procedures performed in studies involving human participants were in accordance with the ethical standards of the institutional and/or national research committee and with the 1964 Helsinki declaration and its later amendments or comparable ethical standards.

References

1. Yuan J, Gu W. Application of miniature pigs as animal models of human diseases in biomedical research. Adv Anim Med. 2011;32(2):108–11.
2. Li J, Wang SL, Zhu XZ, et al. Ultra-micro-pathological observations after injection of methyl viologen into normal parotid glands of miniature pigs. J Mod Stomatol. 2000;14(6):411.
3. Wang SL, Li J, Zhu XZ, et al. Clinical study on the treatment of chronic obstructive parotitis by injecting methyl viologen to cause glandular atrophy. Beijing Stomatol. 2000;8(2):59–63.
4. Li J, Wang SL, Zhu XZ. Histopathological observation of methyl viologen injected into normal parotid glands of miniature pigs. Chin J Stomatol. 1999;34(2):91–3.
5. Xia DS, Qu XM, Tran SD, et al. Histological characteristics following a long-term nitrate-rich diet in miniature pigs with parotid atrophy. Int J Clin Exp Pathol. 2015;8(6):6225–34.
6. Xia DS, Deng DJ, Wang SL. Destruction of parotid glands affects nitrate and nitrite metabolism. J Dent Res. 2003;82(2):101–5.
7. Xia DS, Liu Y, Zhang CM, et al. Antimicrobial effect of acidified nitrate and nitrite on six common oral pathogens *in vitro*. Chin Med J. 2006;119(22):1904–9.
8. Xia DS, Deng DJ, Wang SL. Alterations of nitrate and nitrite content in saliva, serum, and urine in patients with salivary dysfunction. J Oral Pathol Med. 2003;32(2):95–9.

Another Way, Spectacular World

5

Xue Wang

Nitrate is an essential substance for the human body, and the salivary glands are important organs for nitrate enrichment. How, then, do the salivary glands take up and secrete nitrates? Are there any channels in the salivary glands that transport nitrates? If so, what are these channels?

5.1 Nitrate Transporter Sialin

Academician Songlin Wang's research group discovered and reported the cell membrane nitrate transporter sialin for the first time on the basis of a long-term study on salivary glands [1]. Every achievement feels as if it has taken a hundred turns along the road of scientific research. The discovery of the nitrate transporter sialin is the result of more than 10 years of unremitting efforts and unyielding exploration by Songlin Wang's group.

5.1.1 The Discovery of Nitrate Transporter Sialin

5.1.1.1 Raising of Scientific Questions

Why is nitrate so abundant in the saliva, to the extent that it is several times higher than its concentration in the blood? Academician Songlin Wang took this question to heart and found, through literature review, that although there are many reports on nitrates in saliva, the source, transport mechanism, and function of nitrates are basically blank. As a researcher focusing on the salivary gland, he keenly realized that this was a meaningful scientific question and hoped to answer it through research.

X. Wang (✉)
Beijing Stomatological Hospital, Capital Medical University, Beijing, China

S. Wang (ed.), *Inorganic Nitrate and Human Body Homeostasis*,
https://doi.org/10.1007/978-981-95-2824-0_5

Through animal experiments and clinical studies, Academician Songlin Wang's group discovered for the first time that the parotid gland is an important organ for regulating nitrates and maintaining nitrate homeostasis in the body, and it is the main source of nitrates at high concentration in saliva [2, 3]. However, what is the specific mechanism by which salivary glands actively take up and secrete nitrates? Does the human body have a transporter channel for the secretion of nitrates through the salivary glands? With these data and scientific questions in mind, Academician Songlin Wang participated in the second International Symposium on the Salivary Glands held in Pittsburgh, USA, in 2003. He visited the National Institute of Dental and Craniofacial Research (NIDCR) in Professor Bruce Baum's laboratory in the United States of America as a senior visiting scholar. Wang also reported to Professor Xibao Liu, who specialized in cellular electrophysiology, and invited him to join in the research. Prof. Liu readily agreed to cooperate with him and soon found out the working conditions of salivary glands with the patch-clamp technique, which is an electrophysiological measurement method.

5.1.1.2 Discovery of Nitrate Transport Channel

In patch-clamp electrophysiological experiments, it was observed that when the nitrate concentration increased in the extracellular fluid of human submandibular gland cell lines (HSGs), a stable outgoing current change could be detected in the HSG cell membrane. When the pH value of the extracellular fluid decreased from 7.4 to 4, there was a significant increase in the current in the cell membrane. This acid-activated membrane current was usually 3–6 times higher than the spontaneous current, and this current change was independent of extracellular sodium ions. Lowering the extracellular pH also did not produce detectable current changes when chloride, bromide, or gluconic acid concentrations were elevated in the extracellular fluid. When the intracellular nitrate concentration was increased, a change in the internal current could be detected in the cell membrane. This current disappeared quickly, but reducing pH had little effect on it. Some known nonspecific anionic channel inhibitors (e.g., DIDS, NPPB, ECA, CHC) were able to reversibly block the nitrate-induced current changes. In addition, calcium, ATP, cAMP, and other related ions and kinases did not affect nitrate transport by the channel. These results suggest that salivary gland cells have the ability to actively uptake and secrete nitrates, and this current change, dependent on the nitrate concentration, also indicates that HSG cells have specific nitrate channels co-transported with H^+.

After repeated verification, the research group found that the nitrate-induced current changes in the cell membrane are intrinsic and represent a new ion transport channel with distinct characteristics from those previously reported. This suggests that there are previously unreported nitrate transport channels in salivary gland cells. Prof. Liu was excited to get the pre-experimental results and called to inform us that the nitrate-induced current changes in the cell membrane were very stable and reproducible, which was rarely seen in so many electrophysiological measurements of ion channels he had done. The evidence of electrophysiological pre-experiment strengthened the determination of everyone to look further and discover the nitrate transporter channel!

In order to discuss the topic in depth, the research group invited Prof. Liu and his supervisor, Prof. Indu Suresh Ambudkar, to participate in the National Academic Conference on Salivary Glands held in Wuhan in 2005, where Wang's research group gave a special report on the cooperative topic of the nitrate transporter channel. With continuous efforts and repeated attempts, Wang's research group finally confirmed the existence of an inherent and previously unreported nitrate transport channel on the salivary gland cell membrane through patch-clamp electrophysiological experiments. However, by conducting electrophysiological experiments alone, how could the nitrate transport channel be determined? Wang fell into confusion and uncertainty and wondered where to start looking for such a channel.

5.1.1.3 Screening of Nitrate Transport Channel Sialin

The experimental results showed that the nitrate concentration in saliva secreted by the parotid gland was significantly higher than that of the submandibular gland. The parotid and submandibular glands not only differ in their secretions but also perform distinct physiological functions. These differences are primarily attributed to their intrinsic functional genes. Therefore, the research group proposed to first investigate the differences in gene and protein expression between the parotid and submandibular glands. By examining whether the genes encoding nitrate transporter proteins are expressed in these glands at the molecular level, they aimed to provide valuable experimental support for further studies on nitrate transporter channels. After overcoming numerous difficulties, the research group successfully obtained three paired specimens of both glands. Subsequently, gene microarray and protein expression technologies were used to analyze gene expression differences between the parotid and submandibular glands across different individuals. The analysis revealed differences primarily associated with oncogenes, the *Wnt* signaling pathway, anion transporters, mucin synthesis, and other genes. The only functional genes showing differences in ion and protein transport between the two glands were the inwardly rectifying potassium channel KCNJ15, the organic anion transporter proteins SLCO1A2 and SLC13A5, and the chloride channel CFTR [4]. However, none of these ion channels were identified as nitrate transporter protein channels. Subsequently, the research team further searched for the presence of nitrate transporter protein gene expression in the non-differential genes by performing gene expression profiling of the parotid and submandibular glands, with *SLC17A5* (which encodes sialin) and *AQP6* (which encodes aquaporin 6) initially identified as candidate genes. How could their relationship with nitrate transport be established?

At that time, there were very few domestic and international research reports on the nitrate transport system in mammals, with only one relevant study suggesting that *AQP6* exhibited nitrate channel properties in mammals. The expression of *AQP6* in mammalian cells was reported to be localized to intracellular vesicle membranes. However, after adding a green fluorescent protein (GFP) tag to the N-terminal end of *AQP6*, the protein could be expressed on mammalian cell membranes [4]. Patch-clamp studies further revealed that *AQP6* exhibited unique selectivity for nitrate, even in the absence of an acidic environment or Hg^{2+} stimulation. Notably, *AQP6* demonstrated high permeability to nitrate even at a pH of 7.4. Furthermore,

AQP6 could transport both inorganic and organic anions, indicating a broad ion-transporting capacity [5]. In addition, *AQP6* was not inhibited by known anionic channel blockers, such as DIDS, NPPB, DPC, and niflumic acid [4]. When reviewing previous electrophysiological results, it was observed that the active uptake of nitrate by transporter channels in salivary gland cell membranes exhibited pH-dependent characteristics and involved co-transport with H^+ ions. Based on these findings, *AQP6* was not prioritized as a candidate gene for further validation.

Sialin belongs to the anion co-transporter (ACS) subfamily, which is part of the highly conserved major facilitator superfamily (MFS) of membrane transporters found in organisms ranging from plants to higher animals, which also includes plant high-affinity nitrate transporters and drug-resistant proteins. Sialin was initially thought to be a sialic acid/H^+ co-transporter located on the lysosomal membrane [6]. However, at that time, there were no reports on whether sialin was expressed in the human parotid and submandibular glands. To identify the molecules that mediate nitrate/H^+ co-transport in the salivary glands, the research group analyzed the gene expression profiles of the parotid and submandibular glands using gene microarray technology. It was found that the *SLC17A5* gene, which encodes the sialic acid transporter sialin in the MFS superfamily, was highly expressed in both the parotid gland and submandibular gland and functioned as an anion/H^+ co-transporter that can be specifically inhibited by DIDS [1, 7]. Two critical questions emerged that does sialin function as a nitrate transporter and how its transport relationship with nitrate be established? At the time, there was no evidence to support its role in nitrate transport, and further in-depth research seemed impossible. The research team found themselves at a loss, once again facing frustration and uncertainty.

5.1.1.4 Validation of the Nitrate Transport Channel Sialin

It was difficult for the research team to abandon this topic. Although it could be set aside temporarily, giving up was not an option. Therefore, Wang's research group established a method to measure nitrate concentrations inside and outside the cell. They then attempted to silence the *SLC17A5* gene using RNA interference to observe its effect on nitrate transport in salivary gland cells. After long and arduous work, the group discovered that silencing the *SLC17A5* gene significantly decreased the nitrate transport capacity of salivary gland cells, specifically affecting the transport of nitrate without significantly influencing the transport of other ions such as K^+, Na^+, and Cl^-. In addition, silencing of the *SLC17A5* gene resulted in a significant decrease in nitrate membrane currents in HSG cells under both normal and acidic conditions. The positive results of the sialin-nitrate association were simultaneously validated by electrophysiological studies in the United States. The validation of this association provided a preliminary experimental basis for sialin as a nitrate transport channel and rekindled hope for further study. Further in vitro experiments revealed that in HSG cells, sialin transported nitrate in the form of a $2NO_3^-/H^+$ cotransporter, clarifying the mechanism by which sialin transports both nitrate and nitrite.

Subsequently, the research group observed the distribution of sialin throughout the body. Quantitative PCR analysis revealed that sialin was most highly expressed in the parotid gland and brain as well as in major organs such as the liver, kidney, stomach, and spleen. Immunohistochemical staining demonstrated that sialin was specifically expressed in serous acini, with the highest expression observed in the basement membrane. The distribution pattern in acinar cells further supported the anatomical basis of sialin functioning as a nitrate transport channel. Sialin was initially found to be distributed on lysosomal membranes. Different types of mutations can lead to changes in the localization of sialin protein in cells, consequently resulting in altered lysosomal capacity to transport sialic acid. Studies have found that the deletion mutation (268–272) of sialin caused most of the sialin protein to stay in the Golgi apparatus, preventing it from being transported to the lysosomal membrane [8]. There are two pathways through which lysosomal membrane proteins are transported from the Golgi apparatus to the lysosomal membranes: one is the endosomal route, where proteins first pass through the endoplasmic reticulum before reaching the lysosomal membranes; the other is the endocytic route, where proteins reach the lysosomal membrane indirectly through the cell membrane [9]. These observations suggested that lysosomal membranes were not the sole location of sialin protein distribution. Moreover, sequence analysis of sialin did not reveal a classical signal peptide sequence localized to the lysosome [6]. In addition, it was found that sialin was expressed on the cell membrane of primary cultured neurons, without overlap with the specific lysosomal membrane marker lysosomal-associated membrane protein-1 (LAMP-1) [10]. These findings indicated that sialin is not only a lysosomal membrane protein but also has other localizations and broader functions in cells. To further verify the expression location of sialin, the research team performed immunohistochemical double staining. They found that sialin co-stained not only with LAMP-1 but also with cell membrane-specific Na^+–K^+–Cl^- ion channels. This data further provided strong evidence that sialin was present on the cell membrane. These findings also suggested that sialin was not only an ion channel for nitrate transport in salivary gland cells but also an important pathway for regulating nitrate metabolism in the whole body.

However, the most convincing evidence that sialin transports nitrate comes from nitrate loading experiments in *SLC17A5*-mutant humans or from establishing *SLC17A5*-knockout animal models. To this end, Wang's research group simultaneously began searching for patients with *SLC17A5* mutation and developing *SLC17A5* knockout animal models. After overcoming various challenges, fibroblasts were investigated from two patients with sialic acid storage disease (SASD) *SLC17A5* gene mutation (SialinR39C, SialinH183R) from the National Institutes of Health (NIH) of the U.S.A. A series of electrophysiological experiments then further demonstrated the function of sialin in nitrate transport. It was observed that pH-dependent NO_3^- and SA currents were significantly reduced in fibroblasts from the two patients compared with those from healthy volunteers, without altering their current-voltage characteristics. Moreover, transfection of the SialinH183R mutation into fibroblasts from healthy volunteers revealed a significant decrease in the function of sialin to transport NO_3^-. These results further provided strong evidence that sialin functioned as a nitrate transporter. So, what happens when nitrates enter cells?

To address this question, the concentration of nitric oxide in the cells was measured. Repeated experiments demonstrated that nitrates in the cells can be converted into nitric oxide. Such evidence established a sialin-nitrate-nitric oxide correlation, thereby clarifying that cellular translocation of nitrate can play a physiological role via its conversion into nitric oxide [1].

Subsequently, we wrote the manuscript, revised it repeatedly, and experienced many failures such as multiple rejections. Finally, we submitted the paper to *PNAS*, and the editor requested that we revise the manuscript and supplement it with additional in vivo experiments. Although our group acknowledged that additional in vivo experiments would further solidify the latest findings, we faced several practical challenges: (1) SASD patients are phenotypically and genotypically diverse, with very few patients diagnosed worldwide, and most patients are children with severe developmental deficits and poor overall health; (2) adult SASD occurs mainly in the Finnish population, with newborn patients progressively presenting with intellectual disability; (3) rapidly progressive infantile SASD has a more severe onset, and although there is no geographical limitation, embryonic death occurs in early childhood and even in utero. Therefore, the group further verified the previous findings by employing long-accumulated gene transduction technology in miniature pigs. Adenoviral vectors encoding SialinH183R sialin (experimental group) or wild-type sialin (control group) were delivered into parotid ducts, respectively. Our persistence ultimately proved fruitful! There was a significant decrease in nitrate transport function in the parotid glands of miniature pigs transduced with the SialinH183R adenovirus [1], thereby resolving all the questions of the reviewers. The manuscript was accepted following its third revision.

The group's relentless pursuit, hard research work, and unremitting perseverance ultimately yielded a breakthrough: for the first time, the nitrate transport channel sialin was identified on the membrane of mammalian salivary gland epithelial cells, which has unveiled the key channel of nitrate-nitrite-nitric oxide transformation and represents a landmark advancement in both physiology and medicine in the field of stomatology (Fig. 5.1).

5.1.2 Research Significance of Nitrate Transport Channel Sialin

Prof. Lundberg of the Karolinska Institute, Sweden, a renowned expert in physiology and pharmacology, published a special review in *PNAS* at the same time. He concluded that this research represents a crucial first step for nitrate to nitric oxide, which is essential for the physiological metabolism of nitric oxide and its clinical applications. It has important implications for nitric oxide physiology, metabolism, and clinical applications. It also opens up possibilities for further research on the function of this transport channel in human tissues and organs, its role in systemic diseases, and its potential in disease prevention and treatment [11].

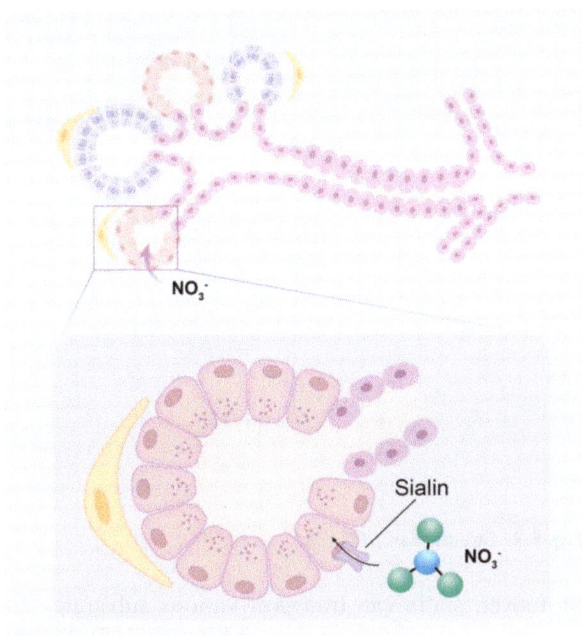

Fig. 5.1 Nitrate transporter sialin

5.2 Functional and Mechanistic Studies of Sialin

Sialin consists of 495 amino acids and contains 12 transmembrane domains. Both its amino-terminal (N-terminal) and carboxy-terminal (C-terminal) ends are located on the cytoplasmic side, enabling its role as an anion transporter (Fig. 5.2) [6]. Mutations in the *SLC17A5* gene can lead to SASD [6]. SASD is an autosomal recessive genetic disease that may cause neurologic, maxillofacial, exocrine gland, liver, pancreas, and kidney developmental abnormalities [6, 12–14]. Based on the *SLCL17A5* mutation genotype and phenotype, SASD is classified into three subtypes: mild (adult sialic acid storage disease, SD), intermediate (severe SD), and severe (infantile sialic acid storage disease, ISSD) [12]. Ninety percent of SD patients harbor the R39C mutation, which is associated with neurological symptoms such as hypotonia, ataxia, and intellectual disability [6, 12]. ISSD results from a variety of mutations, including point mutations, gene deletions, and dislocation insertions; it is mainly characterized by maxillofacial developmental abnormalities, visceral enlargement, and neonatal ascites, with most patients dying within 2 years after birth [12, 13]. The clinical manifestations of severe SD are intermediate between those of the classical SD and ISSD [14]. Sialin was first found to be localized to the lysosomal membrane and transport free sialic acid from the lysosome to the cytoplasm as a H^+/sialic acid cotransporter, hence its designation as sialic acid transporter [6, 15]. Subsequent studies have revealed that sialin also localizes to synaptic vesicles and synapse-like microvesicle membranes, where it functions as a vesicular glutamate/aspartate transporter and transmits neuroexcitatory signals [16].

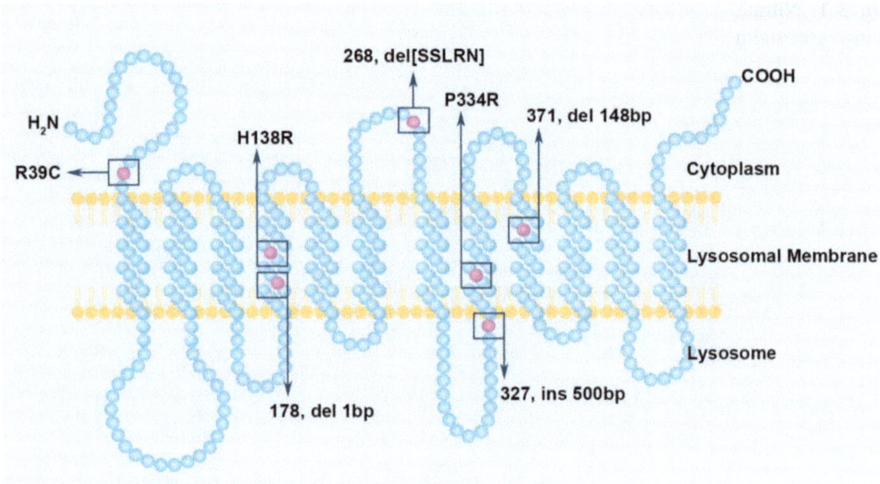

Fig. 5.2 Structure of *SLC17A5*

Moreover, sialin can transport various substrates and exhibits different substrate properties in different organs and tissues [17]. Sialin is closely associated with the salivary gland, and it is highly expressed in the salivary gland during embryonic, juvenile, and adult stages, with dynamic changes in both expression level and expression site [18]. Previous studies by our research group have demonstrated that the parotid gland actively absorbs nitrate from the blood via sialin on the acinar cell membrane and secretes it into saliva. However, what is the physiological significance of the active transport of nitrate in the parotid glands? Is it involved in maintaining normal physiological function of the salivary glands? What specific role does sialin play in this process? Motivated by these scientific questions, and under the leadership of Songlin Wang, our research group continues to move forward along the road of scientific research.

Sialin contains 12 transmembrane structural domains, with both the C- and N-termini located on the cytoplasmic side. Small circles represent amino acid residues; pink circles denote common mutation sites in genes that cause sialic acid storage disease; numbers indicate the positions of amino acids in sialin.

5.2.1 Nitrate-Sialin Feedback Loop Prevent Salivary Gland Radiation Injury

Salivary gland radiation (IR) injury is the most common complication of clinical head and neck tumor radiation therapy [19]. IR-induced salivary gland injury, characterized by symptoms such as excessive salivary secretion, dry mouth, and dysphagia, seriously affects the quality of life of patients; however, there is currently no effective treatment [20]. Songlin Wang's research group confirmed that exogenous

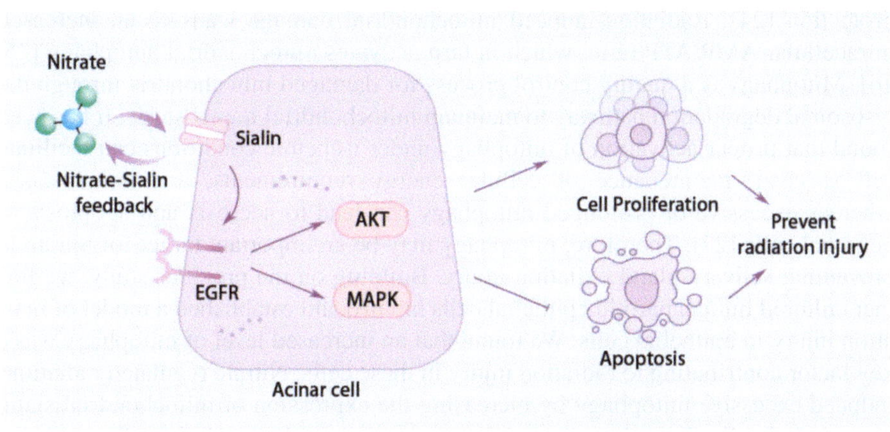

Fig. 5.3 Nitrate-sialin feedback loop prevent salivary gland radiation injury

supplementation of nitrate can effectively prevent parotid radiation injury by establishing a miniature pig parotid radiation injury model. It was found that in the group with nitrate added before radiation, the expression level of sialin in the parotid gland, the salivary flow rate, and body weight increased. Moreover, the parotid gland maintained a normal vesicle structure and vesicle count, and no tissue fibrosis or significant inflammatory reaction was observed. In vitro studies using human parotid epithelial cells found that nitrate interacts with sialin to form a nitrate-sialin feedback loop. Specifically, nitrate promotes the expression of sialin in the acinar cells, and sialin further transports more nitrate into the cells. Furthermore, the EGFR-AKT-MAPK signaling pathway is activated to promote the proliferation of salivary gland acinar cells, inhibit apoptosis, and alleviate radiation damage (Fig. 5.3) [21]. Nitrate is expected to be a new method for the prevention and treatment of salivary gland radiation injury.

Nitrate and sialin form a feedback loop, which activates the EGFR-AKT-MAPK signaling pathway, promotes the proliferation of salivary gland acinus cells, inhibits apoptosis, and reduces radiation damage.

5.2.2 Sialin Regulates Mitophagy

Based on previous research, Wang's research group found that sialin not only functioned as a nitrate transporter but also played a role in regulating mitophagy. Studies have shown that nitrate can protect the normal tissue structure of salivary glands and promote the recovery of salivary gland secretion by reducing radiation-induced reactive oxygen species [22], restoring mitochondrial membrane potential (MMP), and protecting mitochondrial function [23] and other pathways. These results also suggest that mitochondria may be a key subcellular organelle in the protective effect of nitrate. In recent years, more and more studies have shown that mitochondrial DNA is the main target organelle of radiation damage due to the lack of histone protection [24]. Radiation-induced mitochondrial damage leads to an increased intracellular AMP/ATP ratio, which in turn activates mitochondrial autophagy [25, 26]. Mitophagy is a quality control process for damaged mitochondria through the lysosomal degradation pathway to maintain mitochondrial homeostasis. It has been found that proper activation of mitophagy under ischemic conditions can facilitate cell survival, maintenance of cellular energy requirements, and homeostasis, whereas excessive or prolonged mitophagy can lead to necrosis and apoptosis of neuronal cells [27]. Therefore, mitophagy may be an important target for nitrate in preventing salivary gland radiation injury. Building on the previous study, we further cultured human parotid epithelial cells in vitro and established a model of radiation injury in epithelial cells. We found that an increased level of mitophagy was a key factor contributing to radiation injury in these cells. Nitrate regulated radiation-induced excessive mitophagy by increasing the expression of mitochondrial sialin in epithelial cells. To find out how sialin participates in the regulation of mitophagy, the research team then analyzed the cellular localization of sialin and, for the first time, discovered that a portion of sialin was localized to mitochondria, and that nitrate can significantly promote the expression of mitochondrial sialin in epithelial cells after radiation. Sialin, a transmembrane transporter located in the cell membrane of acinar cells [1], is composed of three domains: extracellular, transmembrane, and intracellular. Therefore, the research group hypothesized that sialin may have a transporter-independent function. Subsequently, the protein-binding function of sialin was analyzed by immunoprecipitation coupled with mass spectrometry, and the results suggested that sialin interacts with the mitophagy protein FKBP8. FKBP8 belongs to the FK506-binding protein family of proteins located in the outer membrane of the mitochondria and is capable of recruiting the microtubule-associated protein light chain 3 (LC3) to the damaged mitochondria, causing mitophagy [28, 29]. These findings suggested that nitrate may regulate the radiation-induced increase in mitophagy by promoting the expression of mitochondrial sialin in epithelial cells, interacting with FKBP8, and thus attenuating epithelial cell radiation injury (Fig. 5.4).

Nitrates promote the expression of mitochondrial sialin in epithelial cells and interact with FKBP8 to regulate the excessive increase of mitophagy induced by radiation, thereby alleviating the radiation damage to epithelial cells.

5.2.3 Sialin Involves in Cell Metabolism and Maintains Cell Homeostasis

Nitrates improve mitochondrial function under physiological and pathological conditions [30]. In normal human muscle, nitrates enhance the oxygen effect of mitochondria, thereby improving exercise capacity [30]. In myocardial injury caused by ischemia-reperfusion, nitrates reduce ischemia-reperfusion injury in myocardial tissue by improving mitochondrial function [31]. In addition, in ventricles with abnormal cardiac function due to adriamycin, nitrates reduce adriamycin-induced ventricular functional injury by stabilizing the activity and oxidative

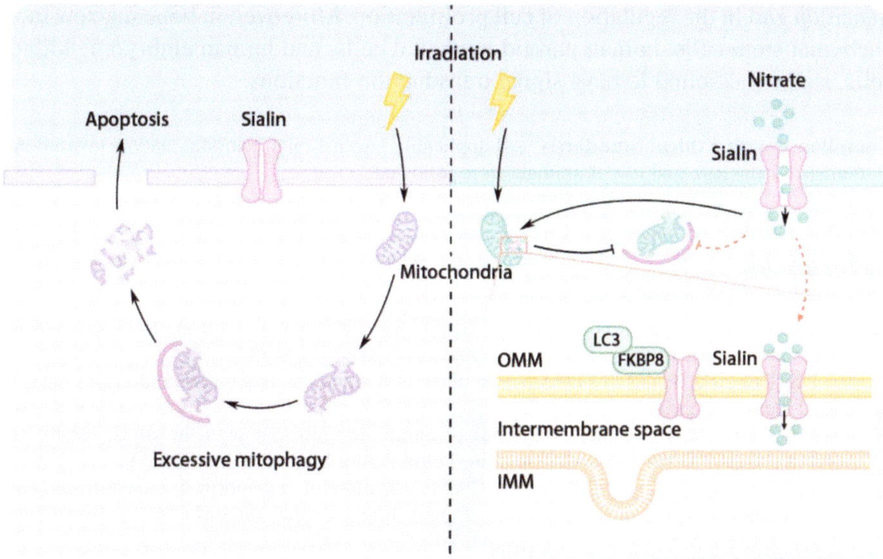

Fig. 5.4 Sialin regulates mitophagy

phosphorylation capacity of mitochondrial complex I and improving mitochondrial function [32]. Deformities and defects of the jaw and joints are extremely common in clinical practice, and the application of mesenchymal stem cells (MSCs) is an excellent "seed cell" for oral and maxillofacial regeneration. The latest findings indicated that nitrates can significantly increase the expression of sialin in MSCs, which plays an indispensable role in various physiological functions as a cell membrane nitrate transporter. In the sialin knockout mouse model, impaired bone formation, reduced osteogenic differentiation potential of bone marrow MSCs, increased senescence level, and functional homeostasis imbalance were observed. Loss- and gain-of-function experiments confirmed that sialin is partially expressed in the mitochondria of MSCs and can enhance osteogenic differentiation potential and reduce senescence level by regulating the mitochondrial function and energy metabolism of MSCs. Wang's research group revealed for the first time that sialin activates mitochondrial metabolism by regulating STAT3 mitochondrial translocation. The results of this study may help to develop new means of targeting and regulating mitochondrial metabolism, delaying cellular aging, and maintaining the functional homeostasis of cells and the organism.

5.2.4 Sialin Involves in Cellular Signal Transduction and Epithelial-Mesenchymal Transformation

Increasing evidence indicates that, in addition to its function as a nitrate transporter, sialin has other functions involved in the maintenance of cellular homeostasis. In thyroid cancer cells, sialin has been found to participate in epithelial-mesenchymal transition and in the regulation of cell proliferation. Moreover, in bone marrow mesenchymal stem cells, human parotid epithelial cells, and human embryonic kidney cells, sialin was found to have signal transduction functions.

Compliance with Ethical Standards All applicable international, national, and/or institutional guidelines for the care and use of animals were followed.

References

1. Qin LZ, Liu XB, Sun QF, et al. Sialin (SLC17A5) functions as a nitrate transporter in the plasma membrane. Proc Natl Acad Sci USA. 2012;109(33):13434–9.
2. Xia DS, Deng DJ, Wang SL. Destruction of parotid glands affects nitrate and nitrite metabolism. J Dent Res. 2003;82(2):101–5.
3. Xia DS, Deng DJ, Wang SL. Alterations of nitrate and nitrite content in saliva, serum, and urine in patients with salivary dysfunction. J Oral Pathol Med. 2003;32(2):95–9.
4. Ikeda M, Beitz E, Kozono D, et al. Characterization of aquaporin-6 as a nitrate channel in mammalian cells. Requirement of pore-lining residue threonine 63. J Biol Chem. 2002;277(42):39873–9.
5. Liu K, Kozono D, Kato Y, et al. Conversion of aquaporin 6 from an anion channel to a water-selective channel by a single amino acid substitution. Proc Natl Acad Sci USA. 2005;102(6):2192–7.
6. Verheijen FW, Verbeek E, Aula N, et al. A new gene, encoding an anion transporter, is mutated in sialic acid storage diseases. Nat Genet. 1999;23(4):462–5.
7. Sun QF, Sun QH, Du J, et al. Differential gene expression profiles of normal human parotid and submandibular glands. Oral Dis. 2008;14(6):500–9.
8. Aula N, Jalanko A, Aula P, et al. Unraveling the molecular pathogenesis of free sialic acid storage disorders: altered targeting of mutant sialin. Mol Genet Metab. 2002;77(1/2):99–107.
9. Yarovaya N, Schot R, Fodero L, et al. Sialin, an anion transporter defective in sialic acid storage diseases, shows highly variable expression in adult mouse brain, and is developmentally regulated. Neurobiol Dis. 2005;19(3):351–65.
10. Aula N, Kopra O, Jalanko A, et al. Sialin expression in the CNS implicates extralysosomal function in neurons. Neurobiol Dis. 2004;15(2):251–61.
11. Lundberg JO. Nitrate transport in salivary glands with implications for NO homeostasis. Proc Natl Acad Sci USA. 2012;109(33):13144–5.
12. Aula N, Salomäki P, Timonen R, et al. The spectrum of SLC17A5-gene mutations resulting in free sialic acid-storage diseases indicates some genotype-phenotype correlation. Am J Hum Genet. 2000;67(4):832–40.
13. Myall NJ, Wreden CC, Wlizla M, et al. G328E and G409E sialin missense mutations similarly impair transport activity, but differentially affect trafficking. Mol Genet Metab. 2007;92(4):371–4.
14. Biancheri R, Verbeek E, Rossi A, et al. An Italian severe Salla disease variant associated with a SLC17A5 mutation earlier described in infantile sialic acid storage disease. Clin Genet. 2002;61(6):443–7.
15. Ruivo R, Anne C, Sagné C, et al. Molecular and cellular basis of lysosomal transmembrane protein dysfunction. Biochim Biophys Acta. 2009;1793(4):636–49.
16. Miyaji T, Echigo N, Hiasa M, et al. Identification of a vesicular aspartate transporter. Proc Natl Acad Sci USA. 2008;105(33):11720–4.
17. Shao ZY, Watanabe S, Christensen R, et al. Synapse location during growth depends on glia location. Cell. 2013;154(2):337–50.
18. Laridon B, Callaerts P, Norga K. Embryonic expression patterns of drosophila ACS family genes related to the human sialin gene. Gene Expr Patterns. 2008;8(4):275–83.

19. Sciubba JJ, Goldenberg D. Oral complications of radiotherapy. Lancet Oncol. 2006;7(2):175–83.
20. Vissink A, Mitchell JB, Baum BJ, et al. Clinical management of salivary gland hypofunction and xerostomia in head-and-neck cancer patients: successes and barriers. Int J Radiat Oncol Biol Phys. 2010;78(4):983–91.
21. Feng XY, Wu ZF, Xu JJ, et al. Dietary nitrate supplementation prevents radiotherapy-induced xerostomia. Elife. 2021;10:e70710.
22. Chang SM, Hu L, Xu YP, et al. Inorganic nitrate alleviates total body irradiation-induced systemic damage by decreasing reactive oxygen species levels. Int J Radiat Oncol Biol Phys. 2019;103(4):945–57.
23. Li SQ, An W, Wang B, et al. Inorganic nitrate alleviates irradiation-induced salivary gland damage by inhibiting pyroptosis. Free Radic Biol Med. 2021;175:130–40.
24. Livingston K, Schlaak RA, Puckett LL, et al. The role of mitochondrial dysfunction in radiation-induced heart disease: from bench to bedside. Front Cardiovasc Med. 2020;7:20.
25. Kam WW, Banati RB. Effects of ionizing radiation on mitochondria. Free Radic Biol Med. 2013;65:607–19.
26. Zhang DY, Wang W, Su XJ, et al. AMPK regulates autophagy by phosphorylating BECN1 at threonine 388. Autophagy. 2016;12(9):1447–59.
27. Hou K, Xu D, Li FY, et al. The progress of neuronal autophagy in cerebral ischemia stroke: mechanisms, roles and research methods. J Neurol Sci. 2019;400:72–82.
28. Edlich F, Lücke C. From cell death to viral replication: the diverse functions of the membrane-associated FKBP38. Curr Opin Pharmacol. 2011;11(4):348–53.
29. Shirane-Kitsuji M, Nakayama KI. Mitochondria: FKBP38 and mitochondrial degradation. Int J Biochem Cell Biol. 2014;51:19–22.
30. Larsen FJ, Schiffer TA, Borniquel S, et al. Dietary inorganic nitrate improves mitochondrial efficiency in humans. Cell Metab. 2011;13(2):149–59.
31. Omar SA, Webb AJ, Lundberg JO, et al. Therapeutic effects of inorganic nitrate and nitrite in cardiovascular and metabolic diseases. J Intern Med. 2016;279(4):315–36.
32. Zhu SG, Kukreja RC, Das A, et al. Dietary nitrate supplementation protects against doxorubicin-induced cardiomyopathy by improving mitochondrial function. J Am Coll Cardiol. 2011;57(21):2181–9.

Part III

Organismic Protection of Nitrates

The high concentration of nitrates in saliva is physiological, necessitating an exploration of nitrates' synergy in both physiological and pathological contexts. Therefore, we conducted a series of thematic research and ended up with several unexpected discoveries.

The Protective Effects of Nitrate on the Digestive System

6

Luyuan Jin, Liang Hu, Shaorong Li, and Zi Yang

The digestive system comprises two primary components: the digestive tract and the digestive glands. The digestive system is composed of the mouth, pharynx, esophagus, stomach, small intestine, and large intestine. Digestive glands can be categorized into two types: major and minor. The salivary glands, liver, and pancreas serve as the primary digestive glands, while additional minor glands are located throughout the walls of the digestive system. Nitrates have demonstrated protective benefits for the stomach, small intestine, colon, pancreas, liver, and salivary glands. Research is currently being conducted on nitrate's physiological protective role and mechanisms in various digestive organs.

6.1 The Protective Effects of Nitrate on the Gastrointestinal Tract

Luyuan Jin

Dietary nitrate consumption occurs in the oral cavity and is absorbed into the circulatory system via the digestive tract. Nitrate's benefits to the human body begin with its direct target organ, the gastrointestinal tract.

(i) The protective role of exogenous nitrates in gastrointestinal injury

As much as 25% of circulating nitrate is actively absorbed by the salivary glands, concentrated roughly tenfold, and subsequently released into the saliva [1, 2]. Commensal bacteria on the tongue's surface, including Veillonella, Staphylococcus aureus, and Staphylococcus epidermidis, can convert salivary nitrates to nitrites.

L. Jin (✉) · L. Hu · S. Li · Z. Yang
Beijing Stomatological Hospital, Capital Medical University, Beijing, China

S. Wang (ed.), *Inorganic Nitrate and Human Body Homeostasis*,
https://doi.org/10.1007/978-981-95-2824-0_6

Nitrites are ingested, enter the stomach, and are converted into nitrogen oxides and nitric oxide (NO) in the acidic milieu [3–5]. The entero-salivary circulation of nitrate is essential for sustaining the dynamic equilibrium of plasma nitrates, nitrites, and NO. Elevated nitrate levels in drinking water can enhance gastric mucosal blood flow, increase mucosal thickness, lower inflammatory cell infiltration, and protect the gastric mucosa from damage induced by dangerous chemicals and ischemia [6, 7]. Nonetheless, prior research has mostly concentrated on the preventive benefits of exogenous nitrate and nitrite administration, while the ability of the organism to actively produce nitrates as a protective inorganic molecule during stress-induced injury remains uncertain.

(ii) Salivary nitrate secretion under stress

Considering that nitrates regulate the oral and gastrointestinal microbiota for protective benefits, does the body actively produce nitrates during acute stress? Based on this scientific theory, we employed bungee jumping as a model for stress. Healthy volunteers were recruited for the study, and their blood pressure, heart rate, respiration, and levels of adrenaline and noradrenaline in the blood were assessed 3 h prior to and immediately following the jump to observe the stress response. Saliva samples were collected 3 h prior to, immediately following, and 1 h post bungee jump to assess the concentration and total quantity of nitrates and nitrites. The findings indicated that the concentration and total quantity of salivary nitrates increased fivefold immediately post jumping and continued to escalate to thirty-threefold 1 hour thereafter. Furthermore, the concentration and total quantity of salivary nitrites reached a peak immediately following the jump, which was about 7.5 times of the level before the bungee jump, and remained consistent for 1 h thereafter (Fig. 6.1). The data suggest that salivary nitrates and nitrites may be actively released under stress conditions.

(iii) The protective role of nitrates in gastric mucosal injury under stress

What is the precise function and mechanism of nitrate secretion by salivary glands under stress? Stress can induce stomach mucosal damage and gastric ulcers. The water-immersion-restraint stress (WIRS) experiment is a traditional animal model for stress-induced gastric ulcers, characterized primarily by a reduction in gastric blood flow. Under stressful settings, the activity of the sympathetic-nervous-vagus-adrenal and hypothalamic-pituitary-adrenal axis is enhanced, alongside the activation and degranulation of mast cells and the reduction of gastric mucus, all of which may contribute to the development of gastric ulcers [8]. To elucidate the function of nitrates in stress-induced gastric mucosal damage, rats were categorized into three groups: a sham-operation control group, a salivary gland duct-ligation group, and a ligation group receiving nitrates in drinking water. Each cohort of animals was submerged in water at 20 °C for a duration of 4 h. The severity of gastric ulcers was assessed by the ulcer index, which measures the average length of gastric ulcers in each group. The findings indicated that gastric mucosal damage was most

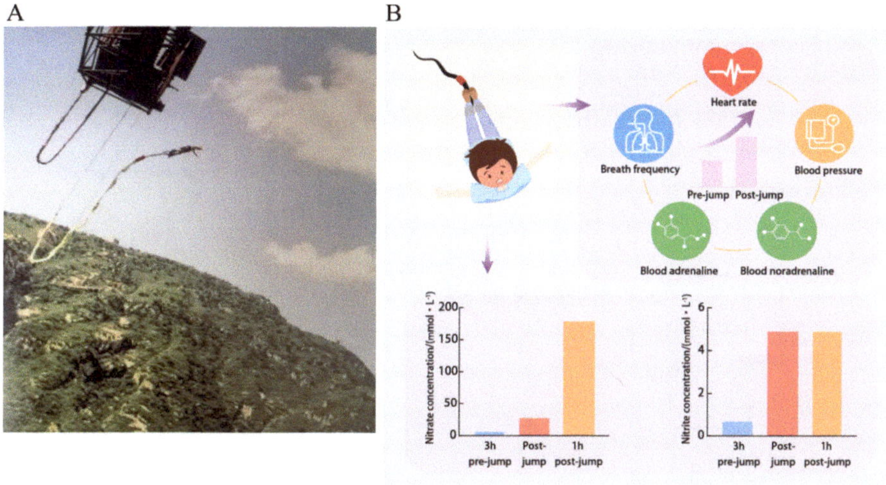

Fig. 6.1 Active nitrate secretion by salivary glands during high-altitude bungee jumping. (**a**) Healthy volunteers undergo bungee jumping at high altitude. (**b**) Physical indices and salivary nitrate and nitrite concentrations before and after bungee the jump (median from 22 volunteers)

severe in the salivary gland duct ligation group, with the ulcer index being 2.41 times greater than that of the control group. Furthermore, NO levels fell by 61% in the gastric cavity relative to the control group, alongside a 25% reduction in gastric mucosal blood flow. Administering nitrates in drinking water prior to stress effectively mitigated gastric mucosal injury, resulting in gastric injury levels comparable to those of the control group. The incorporation of nitrates in the drinking water highly elevated the nitrogen oxide concentration in the gastric cavity, reaching 11 times that of the ligation group. Additionally, the thickness of the gastric mucosal mucus increased, and the gastric mucosal blood flow rose to 1.6 times that of the ligation group. Hence, it was concluded that nitrates can protect the stomach mucosa from stress-induced damage by elevating nitrogen oxide levels in the gastric cavity, enhancing the thickness of gastric mucosal mucus, and improving gastric mucosal blood circulation (Fig. 6.2).

In conclusion, the endogenous secretion of salivary nitrates is adaptively elevated during acute stress. Elevated levels of salivary nitrates and nitrites may provide protection against stress-induced gastric injury via the nitrate-NO pathway. This mechanism enhances gastrointestinal blood flow, fortifies the gastric mucus layer, and diminishes the incidence of gastrointestinal ulcers, erosion, bleeding, and perforations.

(iv) Nitrates could effectively prevent and treat inflammatory bowel disease (Liang Hu)

Intestinal flora establishes and preserves the integrity of the intestinal barrier. Under physiological conditions, intestinal flora and host establish a stable and

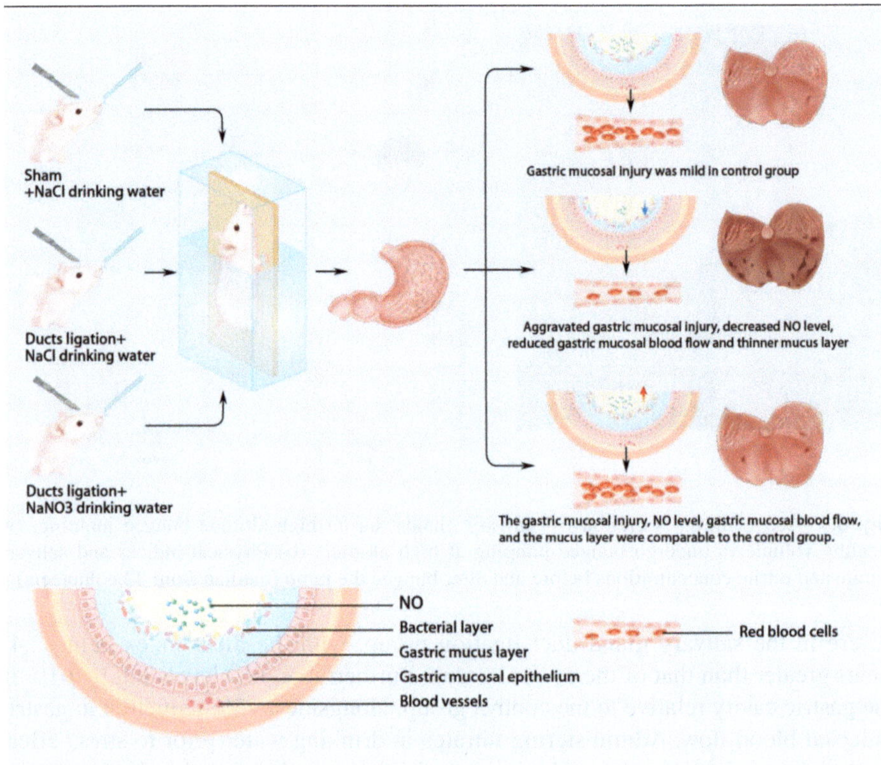

Fig. 6.2 Nitrates effectively protect against gastric mucosal injury under stress

balanced micro-ecosystem. Abnormal intestinal flora impairs intestinal resistance, allowing harmful bacteria to infiltrate and compromising the epithelial barrier. Consequently, several systemic disorders affecting the neurological, cardiovascular, gastrointestinal, immune, and metabolic systems have arisen [9]. Inflammatory bowel disease (IBD) is a collection of nonspecific chronic intestinal inflammatory disorders of uncertain origin, including ulcerative colitis (UC, primarily affecting the rectum and colon) and Crohn's disease (CD, typically occurring in the distal ileum and adjacent colon) [10]. IBD has been associated with immunological problems, bacterial dysbiosis, and genetic predispositions. Previous research indicates that IBD predominantly affects patients in European and American nations, while the incidence of IBD in China remains low [10, 11]. It worth noting that the prevalence of IBD patients has risen year by year, indicating a consistent upward trend.

IBD constitutes the prominent gastrointestinal disease in China. Patients with IBD primarily present with diarrhea, abdominal pain, weight loss, and abdominal masses. The disease typically has a prolonged course, is prone to recurrence, and carries a significant risk for colon cancer [12, 13]. IBD is primarily managed with medication, with 5-aminosalicylic acid (5-ASA) serving as the first-line treatment for maintaining mild to moderate ulcerative colitis [14]. Glucocorticoid medications, such as budesonide and beclomethasone dipropionate, are utilized in the treatment of moderate to severe IBD. As the incidence of IBD rises annually, the pharmacological agents employed in clinical treatment may lead to various adverse side effects [15, 16]. Therefore, the management of IBD requires additional investigation.

The incidence of IBD is closely related to diet, with a positive association observed between increased intake of refined sugar, meat, and fat, and a decreased intake of fiber [17]. Nutritional therapy can sustain therapeutic effects and alleviate symptoms. Enteral nutrition therapy (EN) consists of two primary forms. First, a basic diet includes nutrients in simple forms, such as amino acids, monosaccharides, oligosaccharides, and medium-chain triglycerides. Second, an aggregative diet comprises whole protein and starchy hydrolyzed carbohydrates, which can effectively alleviate IBD symptoms [18, 19]. Epigallocatechin gallate, the primary component of tea polyphenols, exhibits bio-functions such as anti-inflammation and oxidative stress injury reduction of up-regulating colonic microbiota, specifically Akkermansia, and enhancing short-chain fatty acids production [20]. The IL23/IL17 signaling pathway is significant in the pathogenesis of IBD, and its inhibition may represent a viable therapeutic approach for IBD [21].

The research group led by Academician Songlin Wang developed the IBD mice model by administering dextran sulfate sodium (DSS) in drinking water. Subsequently, exogenous nitrates were employed to examine the treatment effectiveness of IBD colitis. Experimental results demonstrated that nitrate supplementation significantly reduced the disease activity index (DAI) in the mouse model. DSS may enhance Caspase signaling in the intestinal epithelium, leading to the death of intestinal epithelial cells, an effect that is counteracted by nitrates. Nitrates may down-regulate the expression of IL-17 and MMP2, hence reducing the inflammatory response generated by DSS.

Our results indicate that nitrates may reduce the inflammatory cytokines levels. In addition, nitrates may down-regulate the abundance of various pathogenic bacteria, including *Bacteroidales_S24-7_group_unidentified*, *Bacteroides*, and *Prevotellaceae_UCG-001* et al., while simultaneously up-regulating the abundance of *Lactobacillus* and *Ruminococcaceae_UCG-014* [22]. Nitrate supplementation may mitigate IBD colitis in mice by reducing inflammatory and apoptotic levels in the intestinal epithelium through the regulation of the colonic microbiome (Fig. 6.3).

6.2 The Protection of Nitrate on Liver

Shaorong Li

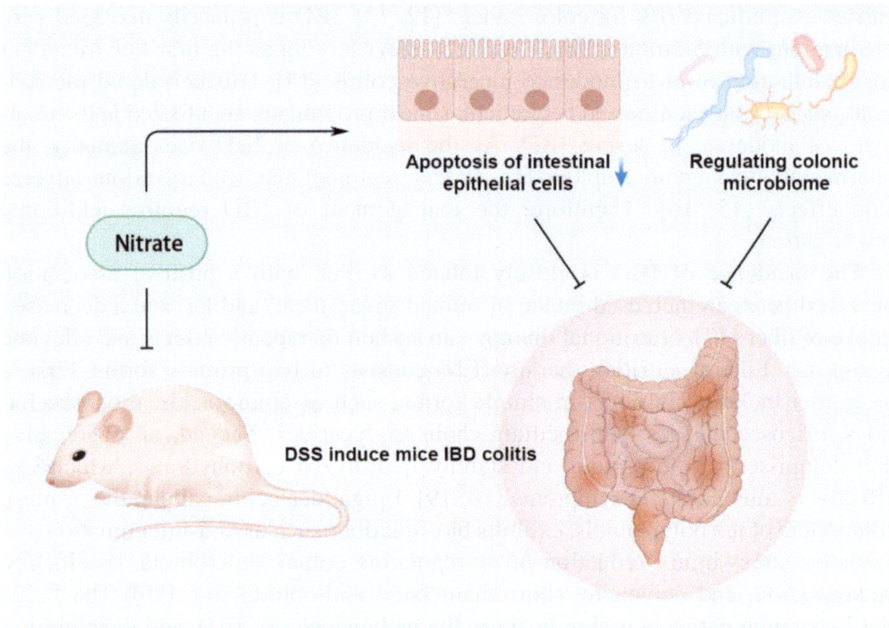

Fig. 6.3 Nitrate oral administration significantly reduces inflammatory colonic diseases

6.2.1 Non-alcoholic Fatty Liver Disease

In recent years, alterations in lifestyle and diet have led to an increased prevalence of metabolic syndrome. Inflammatory-related disorders, particularly non-alcoholic fatty liver disease (NAFLD), have become one of the most common liver conditions, increasingly impacting younger populations [23]. In the United States, 25% of the population suffers from NAFLD, which is currently the second biggest cause of liver transplants [24]. NAFLD can be classified into steatosis, non-alcoholic steatohepatitis, liver fibrosis, cirrhosis, and primary liver cancer based on its progression [25]. Non-alcoholic steatohepatitis is a critical phase in the disease's course, serving as a necessary intermediary between steatosis and the development of liver fibrosis and cirrhosis. Upon advancing to non-alcoholic steatohepatitis, the disease trajectory becomes almost irreversible [26]. Currently, there are no efficacious pharmacological interventions in clinical practice for the prevention and treatment of NAFLD.

6.2.1.1 Nitrates Re-balance the Homeostasis of Hepatic Macrophages for NAFLD Prevention and Treatment

Utilizing a methionine-choline-deficient diet (MCD) and a choline-deficient high-fat diet (CDHFD) to create a NAFLD mouse model, along with prior administration of nitrate water, has demonstrated that oral nitrate intake can impede the advancement of diet-induced NAFLD. This is demonstrated by the reduction of hepatic inflammation and steatosis, decreased activities of alanine aminotransferase (ALT) and aspartate aminotransferase (AST) (Fig. 6.4), and lower levels of fibrosis. Mechanistic investigations indicate that the protective effects of nitrates primarily function through the modulation of hepatic bone marrow-derived macrophages (MoMFs), particularly by decreasing the overall MoMF levels. Subsequent investigation indicates that nitrates increase the ratio of anti-inflammatory MoMFs (identified by *Arg1*) in the liver, concurrently decreasing the ratio of pro-inflammatory MoMFs (identified by *Inos*) (Fig. 6.4). Experiments were conducted to identify the target cells affected by nitrate action, utilizing oleic acid (OA) to induce hepatic cell steatosis. The results indicated that the addition of nitrates did not reduce the extent of steatosis. In MoMFs collected from isolated and cultured mice, factors were added individually to facilitate their differentiation into anti-inflammatory and pro-inflammatory MoMFs. In an in vitro study, the addition of nitrates was observed to enhance the proportion of anti-inflammatory MoMFs and stimulate IL-10 secretion, while simultaneously reducing the proportion of pro-inflammatory MoMFs, thereby inhibiting TNF-α secretion (Fig. 6.4). This demonstrates that in the NAFLD model, nitrates primarily influence hepatic MoMFs rather than hepatic cells.

6.2.1.2 Nitrates Regulate the Hemostasis of Hepatic Macrophages by Sialin

Prior studies have clarified that the effectiveness of nitrates in humans mostly relies on the nitrate-nitrite-NO pathway, facilitated by NO. [27] Nitrates are converted into nitrites by oral and gastrointestinal microbiota throughout this process. Nitrite, known for its significant instability, can be transformed into NO by an enzymatic mechanism [28]. Recent results indicate that the nitrite reduction process mediated by xanthine oxidase might effectively impair the activity of activated macrophages, potentially influencing inflammatory responses. Analysis of its mechanism reveals that nitrites alone do not modify intracellular NO levels, indicating a direct biological effect of nitrites on activated macrophages, independent of NO production [29]. Certain researchers have revealed that nitrates inhibit both acute and chronic inflammation by modulating neutrophil recruitment, thereby partially mitigating the inflammatory condition of atherosclerotic plaques. This finding may present clinical significance [30]. The academics contend that the aforementioned effects cannot be solely attributed to the nitrate-nitrite-NO pathway, due to the brief physiological half-life of NO, which consequently exerts a restricted influence [31]. The methods via which dietary nitrates influence the immune system are a critical subject requiring additional research.

The sterile in vitro cell culture environment suggests the absence of the nitrate-NO pathway. Thus, nitrate's in vitro actions may engage other systems. To clarify whether nitrates can play a role independent of NO, broad-spectrum antibiotics (ABX) are employed to develop largely germ-free mouse models [32]. Moreover, it was noted that despite the attenuation of NO effects, nitrates might still influence macrophages to avert NAFLD (Fig. 6.5), suggesting the existence of potential novel mechanisms. Currently, the identification of the nitrate transporter sialin's potential involvement, the extraction of *Slc17a5* sgRNA from two-cell embryo mice [33] for

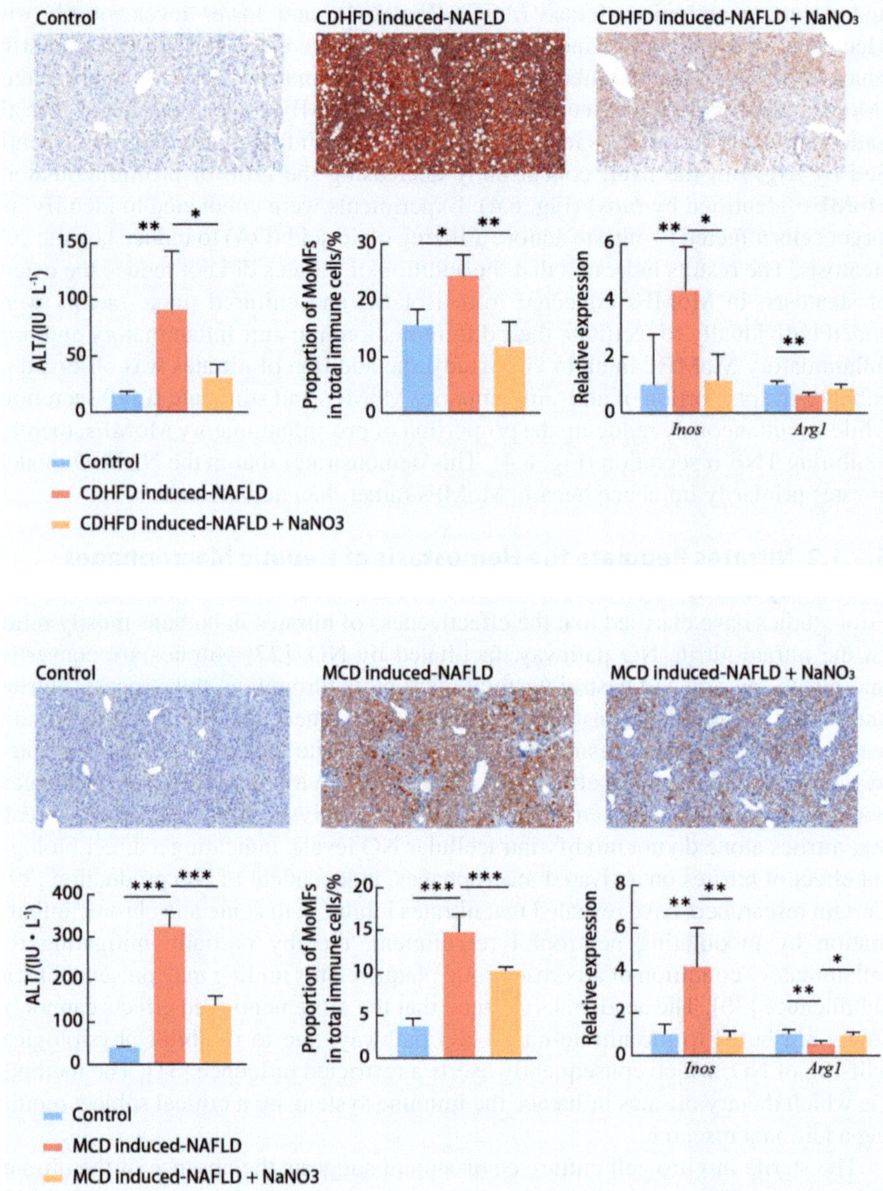

Fig. 6.4 Nitrates re-balance hepatic macrophage for NAFLD prevention and treatment

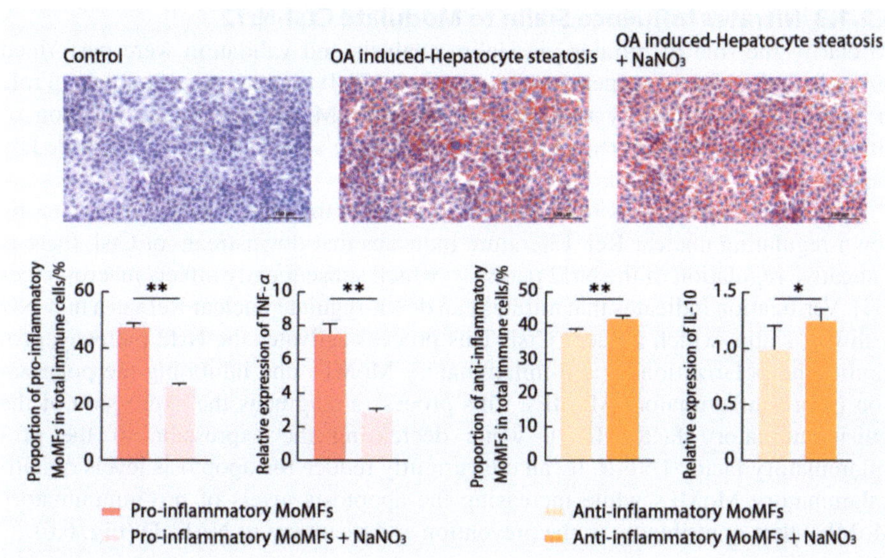

Fig. 6.4 (continued)

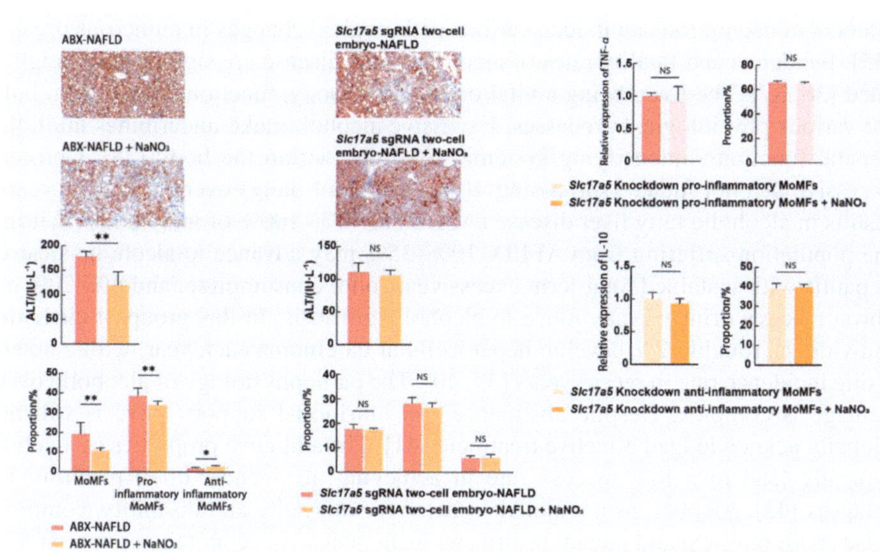

Fig. 6.5 Nitrates re-balance hepatic macrophage homeostasis by sialin to prevent and treat NAFLD

bone marrow reconstruction, and the execution of in vitro cell experiments to validate the knockdown of *Slc17a5* reveal that the inhibition of sialin's function significantly reduces the regulatory effect of nitrates on macrophages NAFLD prevention (Fig. 6.5). This highlights that, in addition to the traditional NO pathway, nitrates can directly influence macrophages to avert NAFLD via sialin.

6.2.1.3 Nitrates Influence Sialin to Modulate Ctsl-Nrf2

To clarify the role of nitrates via sialin, analysis and validation were performed through transcriptome sequencing. Cathepsin L (Ctsl) is shown to play a crucial role in both anti-inflammatory and pro-inflammatory MoMFs, with the addition of nitrates resulting in a reduction of Ctsl levels. Nitrate's effect is severely impeded by the over expression of Ctsl.

CUT&Tag experiments revealed and confirmed that sialin may reduce Ctsl by down-regulating nuclear Rel. Literature indicates that downstream of Ctsl, there is a negative regulation of the Nrf2 pathway, which subsequently affects macrophages [34]. Verification indicates that nitrates can down-regulate nuclear Rel via a non-NO pathway, sialin, which reduces Ctsl. This process activates the Nrf2 pathway, promoting the polarization of anti-inflammatory MoMFs and inhibiting the polarization of pro-inflammatory MoMFs. This process strengthens the expression of the anti-inflammatory factor IL-10 while decreasing the expression of the pro-inflammatory factor TNF-α. It can concurrently reduce the apoptosis levels of anti-inflammatory MoMFs while increasing the apoptosis levels of pro-inflammatory MoMFs, thus contributing to the prevention and treatment of NAFLD (Fig. 6.6).

6.2.2 Alcoholic Fatty Liver Disease

Alcohol consumption can induce various pathological changes in numerous organs [35]. Incidence and fatality rates of alcoholic liver disease are significantly heightened [36, 37]. The liver, being a vital organ in the body, functions as a central hub for various physiological processes. Excessive alcohol intake undermines multiple hepatic functions, intensifying systemic reactions within the body [38]. Chronic excessive alcohol intake (surpassing 40 g of ethanol daily) over months or years results in alcoholic fatty liver disease (AFLD) in 90%–100% of individuals. Within the population suffering from AFLD, 10%–35% may advance to alcoholic steatohepatitis with sustained long-term excessive alcohol consumption, and 8%–20% of chronic heavy drinkers may move to alcoholic cirrhosis. In this group of cirrhotic individuals, roughly 2% develop hepatocellular carcinoma each year, with a noted rising incidence rate in recent years [39, 40]. The pathophysiology of alcoholic liver damage is complex. Despite abstinence and nutritional support, there is still no globally acknowledged effective treatment [41]. The addictive properties of alcohol typically lead to a low success rate in achieving abstinence, often resulting in relapses [42]. Alcohol, as a major risk factor for mortality and disability, compromises both physical and mental health and incurs negative economic repercussions. It is closely associated with elevated risks of public accidents, diminished productivity, and increased rates of crime and violence [43, 44]. Preventing and treating alcohol-related liver problems is crucial for societal security and economic progress, necessitating additional comprehensive study.

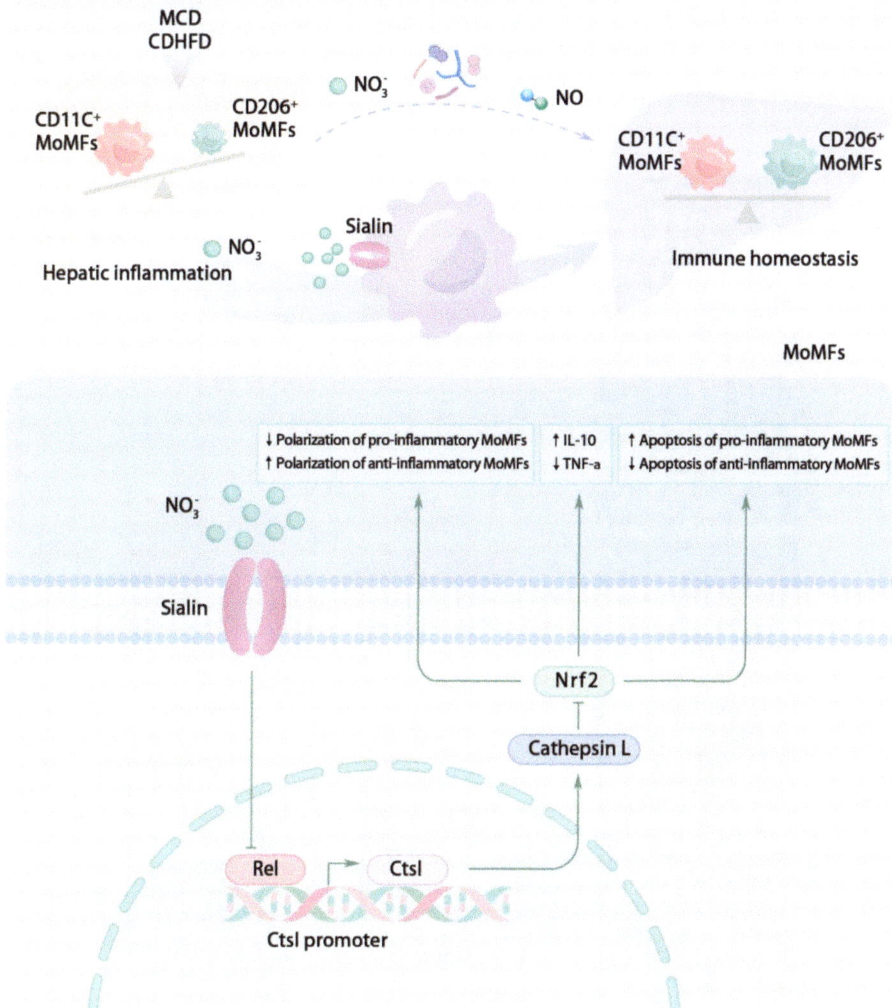

Fig. 6.6 Nitrates modulate hepatic macrophage homeostasis via the Sialin-Ctsl-Nrf2 pathway, contributing to the prevention and treatment of NAFLD

6.2.2.1 Nitrates Prevent AFLD

Two animal models were established: an acute alcohol exposure model using single-dose ethanol gavage, and a classic alcohol-induced fatty liver disease model as defined by the National Institute on Alcohol Abuse and Alcoholism (NIAAA), which involves specialized diets and ethanol gavage. Previous administration of nitrates significantly reduces alcohol-induced elevations in serum ALT, AST, triglycerides (TG), total cholesterol (TC), and free fatty acid levels. This intervention significantly lowers hepatic inflammatory infiltration and the levels of steatosis (Fig. 6.7).

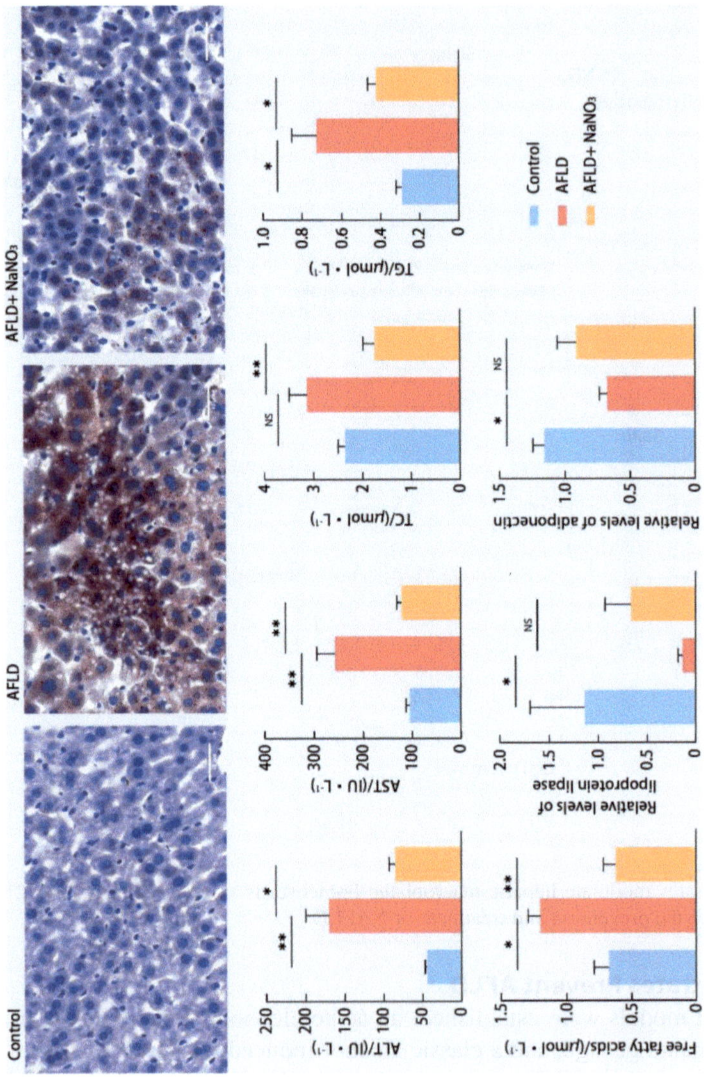

Fig. 6.7 Nitrates prevent and treat AFLD

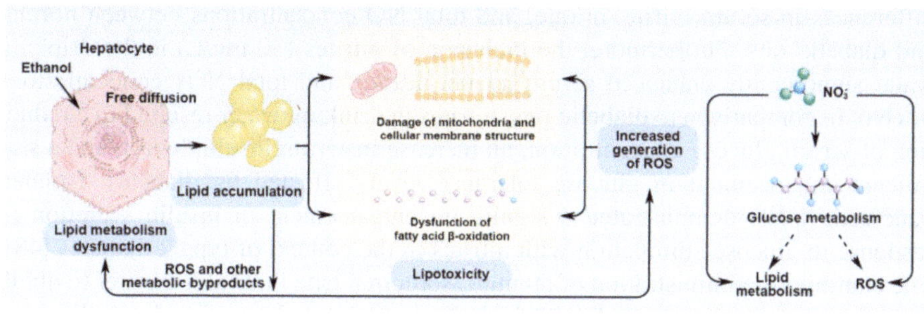

Fig. 6.8 Nitrates alleviate AFLD by regulating glucose and lipid metabolism

6.2.2.2 Nitrates Function Through Lipid Metabolism Regulation

The investigation into its mechanism indicates that nitrates can modulate essential enzymes, including lipoprotein lipase, adiponectin, and fatty acid β-oxidation, thereby facilitating hepatic lipid metabolism enhancement and AFLD prevention (Fig. 6.8).

6.3 The Protective Role of Nitrates in Other Organs of the Digestive System

Zi Yang

NO molecules are crucial in the progression and management of diabetes. Deficiency in endothelial nitric oxide synthase (eNOS) can lead to metabolic dysfunctions, including dyslipidemia and impaired glucose tolerance [45]. In comparison to the conventional diet, a low nitrite/nitrate diet administered to mice for 3 months led to visceral obesity, dyslipidemia, and impaired glucose tolerance. Furthermore, a low nitrite/nitrate diet for 18 months resulted in notable increases in body weight, hypertension, insulin resistance, and impaired endothelium-dependent vasodilation in response to acetylcholine. Additionally, a low nitrite/nitrate diet over 22 months was associated with a significant increase in mortality from cardiovascular disease [46]. Dietary nitrite/nitrate play a crucial role in pancreatic metabolic function and may be associated with NO molecules.

In animal experiments, intraperitoneal nitrite injection significantly elevated plasma nitrite concentrations within the physiological range. Total pancreatic blood flow, blood glucose, and glucose tolerance remained unchanged; however, pancreatic islet blood flow increased by 50%, and circulating insulin levels rose by 30%. An in vitro study on isolated islet cells demonstrated that, under basal conditions, nitrites enhanced insulin secretion from rat pancreatic islet β-cells in a dose-dependent manner through the NO/GC/cGMP pathway across a broad concentration range (0.1–100 mmol/L) [47]. The creation of a type 2 diabetic rat model through a high-fat diet and low-dose streptozotocin indicated that there were no significant differences in serum nitrite, nitrate, and total NO concentrations between normal and diabetic rats. Furthermore, the inclusion of nitrites (50 mg/L) in the drinking water significantly enhanced serum nitrite, nitrate, and total NOx concentrations in vivo. In comparison to diabetic rats, nitrites in drinking water resulted in a reduction of serum glucose concentration, an increase in serum insulin levels, and a significant enhancement in glucose tolerance levels affected by diabetes. Isolated pancreatic islets demonstrated a significant enhancement in insulin secretion in response to glucose stimulation with nitrites in the context of type 2 diabetes [48]. The continuous administration of 50 mg/L sodium nitrite in drinking water to db/db diabetic mice over a period of 4 weeks led to reduced weight gain, elevated fasting blood glucose levels, and decreased insulin levels in db/db mice. Additionally, nitrites enhanced insulin function by restoring the signaling translocation of glucose transporters 4 (GLUT4) in vitro [49]. Nitrate supplementation largely strengthened glucose tolerance and catalase activity while reducing gluconeogenesis, fasting glucose, and insulin levels in diabetic rats. However, it did not significantly affect glucose-stimulated insulin secretion (GSIS), insulin levels, or glycated hemoglobin (HbA1c). Nitrate supplementation in diabetic rats resulted in elevated GLUT4 levels in the adipose tissue of flounder muscle and epididymis [50]. In an animal model of hyperglycemia, dietary nitrate mitigated increases in systolic blood pressure and serum glucose, enhanced glucose tolerance, and restored dyslipidemia [51].

Compliance with Ethical Standards All applicable international, national, and/or institutional guidelines for the care and use of animals were followed.

All procedures performed in studies involving human participants were in accordance with the ethical standards of the institutional and/or national research committee and with the 1964 Helsinki declaration and its later amendments or comparable ethical standards.

References

1. Pannala AS, Mani AR, Spencer JP, et al. The effect of dietary nitrate on salivary, plasma, and urinary nitrate metabolism in humans. Free Radic Biol Med. 2003;34(5):576–84.
2. GM MK, Duncan CW, Leifert C, et al. Dietary nitrate in man: friend or foe? Br J Nutr. 1999;81(5):349–58.
3. Lundberg JO, Weitzberg E, Gladwin MT. The nitrate-nitrite–nitric oxide pathway in physiology and therapeutics. Nat Rev Drug Discov. 2008;7(2):156–67.
4. Larsen FJ, Ekblom B, Sahlin K, et al. Effects of dietary nitrate on blood pressure in healthy volunteers. N Engl J Med. 2006;355(26):2792–3.
5. Benjamin NO, Driscoll F, Dougall H, et al. Stomach NO synthesis. Nature. 1994;368(6471):502.
6. Petersson J, Phillipson M, Jansson EA, et al. Dietary nitrate increases gastric mucosal blood flow and mucosal defense. Am J Physiol Gastrointest Liver Physiol. 2007;292(3):G718–24.
7. Jansson EA, Petersson J, Reinders C, et al. Protection from nonsteroidal anti-inflammatory drug (NSAID)-induced gastric ulcers by dietary nitrate. Free Radic Biol Med. 2007;42(4):510–8.
8. Bhatia V, Tandon RK. Stress and the gastrointestinal tract. J Gastroenterol Hepatol. 2005;20(3):332–9.
9. O'Hara AM, Shanahan F. The gut flora as a forgotten organ. EMBO Rep. 2006;7(7):688–93.
10. Yu YR, Rodriguez JR. Clinical presentation of Crohn's, ulcerative colitis, and indeterminate colitis: symptoms, extraintestinal manifestations, and disease phenotypes. Semin Pediatr Surg. 2017;26(6):349–55.

11. Hamdeh S, Micic D, Hanauer S. Drug-induced colitis. Clin Gastroenterol Hepatol. 2021;19(9):1759–79.
12. Ng WK, Wong SH, Ng SC. Changing epidemiological trends of inflammatory bowel disease in Asia. Intest Res. 2016;14(2):111–9.
13. Ran Z, Wu KC, Matsuoka K, et al. Asian Organization for Crohn's and colitis and Asia Pacific Association of Gastroenterology practice recommendations for medical management and monitoring of inflammatory bowel disease in Asia. J Gastroenterol Hepatol. 2021;36(3):637–45.
14. Le Berre C, Roda G, Nedeljkovic Protic M, et al. Modern use of 5-aminosalicylic acid compounds for ulcerative colitis. Expert Opin Biol Ther. 2020;20(4):363–78.
15. Bruscoli S, Febo M, Riccardi C, et al. Glucocorticoid therapy in inflammatory bowel disease: mechanisms and clinical practice. Front Immunol. 2021;12:691480.
16. Pithadia AB, Jain S. Treatment of inflammatory bowel disease (IBD). Pharmacol Rep. 2011;63(3):629–42.
17. Hou JK, Abraham B, El-Serag H. Dietary intake and risk of developing inflammatory bowel disease: a systematic review of the literature. Am J Gastroenterol. 2011;106(4):563–73.
18. Smith PA. Nutritional therapy for active Crohn's disease. World J Gastroenterol. 2008;14(27):4420–3.
19. Durchschein F, Petritsch W, Hammer HF. Diet therapy for inflammatory bowel diseases: the established and the new. World J Gastroenterol. 2016;22(7):2179–94.
20. Wu ZH, Huang SM, Li TT, et al. Gut microbiota from green tea polyphenol-dosed mice improves intestinal epithelial homeostasis and ameliorates experimental colitis. Microbiome. 2021;9(1):184.
21. Noviello D, Mager R, Roda G, et al. The IL23-IL17 immune axis in the treatment of ulcerative colitis: successes, defeats, and ongoing challenges. Front Immunol. 2021;12:611256.
22. Hu L, Jin LY, Xia DS, et al. Nitrate ameliorates dextran sodium sulfate-induced colitis by regulating the homeostasis of the intestinal microbiota. Free Radic Bio Med. 2020;152:609–21.
23. Borrelli A, Bonelli P, Tuccillo FM, et al. Role of gut microbiota and oxidative stress in the progression of non-alcoholic fatty liver disease to hepatocarcinoma: current and innovative therapeutic approaches. Redox Biol. 2018;15:467–79.
24. Diehl AM, Day C. Cause, pathogenesis, and treatment of nonalcoholic steatohepatitis. N Engl J Med. 2017;377(21):2063–72.
25. Despres JP, Lemieux I. Abdominal obesity and metabolic syndrome. Nature. 2006;444(7121):881–7.
26. Suzuki A, Diehl AM. Nonalcoholic steatohepatitis. Annu Rev Med. 2017;68:85–98.
27. Witter JP, Gatley SJ, Balish E. Distribution of nitrogen-13 from labeled nitrate (13No3-) in humans and rats. Science. 1979;204(4391):411–3.
28. Bryan NS, Ivy JL. Inorganic nitrite and nitrate: evidence to support consideration as dietary nutrients. Nutr Res. 2015;35(8):643–54.
29. Yang T, Peleli M, Zollbrecht C, et al. Inorganic nitrite attenuates NADPH oxidase-derived superoxide generation in activated macrophages via a nitric oxide-dependent mechanism. Free Radic Biol Med. 2015;83:159–66.
30. Khambata RS, Ghosh SM, Rathod KS, et al. Anti-inflammatory actions of inorganic nitrate stabilize the atherosclerotic plaque. Proc Natl Acad Sci USA. 2017;114(4):E550–9.
31. Yang T, Zhang XM, Tarnawski L, et al. Dietary nitrate attenuates renal ischemia-reperfusion injuries by modulation of immune responses and reduction of oxidative stress. Redox Biol. 2017;13:320–30.
32. Shi JX, Wang YF, He J, et al. Intestinal microbiota contributes to colonic epithelial changes in simulated microgravity mouse model. FASEB J. 2017;31(8):3695–709.
33. Wu Y, Zhang J, Peng B, et al. Generating viable mice with heritable embryonically lethal mutations using the CRISPR-Cas9 system in two-cell embryos. Nat Commun. 2019;10(1):2883.
34. Sato T, Yamashina S, Izumi K, et al. Cathepsin L-deficiency enhances liver regeneration after partial hepatectomy. Life Sci. 2019;221:293–300.

35. Mehta G, Sheron N. No safe level of alcohol consumption—implications for global health. J Hepatol. 2019;70(4):587–9.
36. Sarin SK, Kumar M, Eslam M, et al. Liver diseases in the Asia-Pacific region: a Lancet Gastroenterology & Hepatology Commission. Lancet Gastroenterol Hepatol. 2020;5(2):167–228.
37. GBD 2016 Causes of Death Collaborators. Global, regional, and national age-sex specific mortality for 264 causes of death, 1980–2016: a systematic analysis for the global burden of disease study 2016. Lancet. 2017;390(10100):1151–210.
38. Trefts E, Gannon M, Wasserman DH. The liver. Curr Biol. 2017;27(21):R1147–51.
39. Seitz HK, Bataller R, Cortez-Pinto H, et al. Alcoholic liver disease. Nat Rev Dis Primers. 2018;4(1):16.
40. Bajaj JS. Alcohol, liver disease and the gut microbiota. Nat Rev Gastroenterol Hepatol. 2019;16(4):235–46.
41. Singal AK, Mathurin P. Diagnosis and treatment of alcohol-associated liver disease: a review. JAMA. 2021;326(2):165–76.
42. Albillos A, De Gottardi A, Rescigno M. The gut-liver axis in liver disease: pathophysiological basis for therapy. J Hepatol. 2020;72(3):558–77.
43. Witkiewitz K, Litten RZ, Leggio L. Advances in the science and treatment of alcohol use disorder. Sci Adv. 2019;5(9):eaax4043.
44. Aldridge RW, Story A, Hwang SW, et al. Morbidity and mortality in homeless individuals, prisoners, sex workers, and individuals with substance use disorders in high-income countries: a systematic review and meta-analysis. Lancet. 2018;391(10117):241–50.
45. Cook S, Hugli O, Egli M, et al. Clustering of cardiovascular risk factors mimicking the human metabolic syndrome X in eNOS null mice. Swiss Med Wkly. 2003;133(25/26):360–3.
46. Kina-Tanada M, Sakanashi M, Tanimoto A, et al. Long-term dietary nitrite and nitrate deficiency causes the metabolic syndrome, endothelial dysfunction and cardiovascular death in mice. Diabetologia. 2017;60(6):1138–51.
47. Nyström T, Ortsäter H, Huang Z, et al. Inorganic nitrite stimulates pancreatic islet blood flow and insulin secretion. Free Radic Biol Med. 2012;53(5):1017–23.
48. Gheibi S, Bakhtizrzadeh F, Jeddi S, et al. Nitrite increases glucose-stimulated insulin secretion and islet insulin content in obese type 2 diabetic male rats. Nitric Oxide. 2017;64:39–51.
49. Jiang H, Torregrossa AC, Potts A, et al. Dietary nitrite improves insulin signaling through GLUT4 translocation. Free Radic Biol Med. 2014;67:51–7.
50. Gheibi S, Jeddi S, Carlström M, et al. Effects of long-term nitrate supplementation on carbohydrate metabolism, lipid profiles, oxidative stress, and inflammation in male obese type 2 diabetic rats. Nitric Oxide. 2018;75:27–41.
51. Kalifi S, Rahimipour A, Jeddi S, et al. Dietary nitrate improves glucose tolerance and lipid profile in an animal model of hyperglycemia. Nitric Oxide. 2015;44:24–30.

Protective Effects on the Cardiovascular System

7

Zhichao Chang, Jianyu Gu, and Xiaoyu Li

Cardiovascular disease (CVD) is extremely common worldwide, with hypertension, ischemic heart disease, and stroke being the primary causes of death from CVD. The high prevalence imposes a major strain on healthcare system. As a result, it is critical to both prevent cardiovascular illnesses and avert significant adverse cardiovascular events. Our study discovered that nitrates have a number of favorable physiological benefits in the prevention and treatment of cardiovascular disease.

7.1 Progress in the Study of Nitrated Fatty Acids

Zhichao Chang

Since the 1986 discovery of endogenous NO production and its physiological effects on vasodilation, the use of nitrates in cardiovascular diseases has persisted [1]. Nevertheless, NO, a messenger molecule, is extremely soluble and has a short duration of action. Consequently, there has been a shift in research focus to upstream substances that regulate NO production. Nitrated fatty acids (NO2-FA), which are NO derivatives, have progressively garnered the attention of a greater number of researchers due to their reliance on the production of inflammatory responses (Fig. 7.1). Nitrated fatty acids may modulate inflammatory responses by participating in a variety of signaling pathways that are associated with anti-inflammatory effects.

Z. Chang · X. Li (✉)
Beijing Stomatological Hospital, Capital Medical University, Beijing, China
e-mail: xiaoyu9449@ccmu.edu.cn

J. Gu
Stomatological Hospital of Shandong University, Jinan, Shandong, China

S. Wang (ed.), *Inorganic Nitrate and Human Body Homeostasis*,
https://doi.org/10.1007/978-981-95-2824-0_7

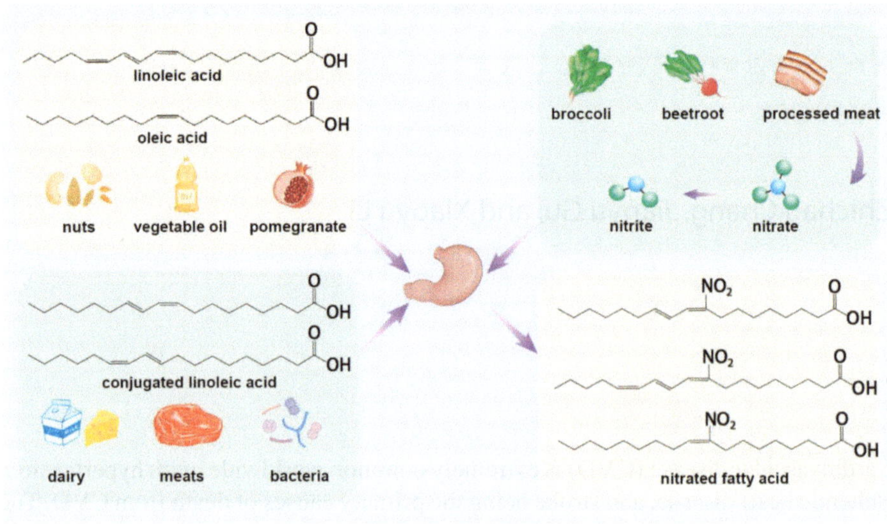

Fig. 7.1 Sources of nitro-fatty acids

7.1.1 Generation of Nitrated Fatty Acids

NO, produced metabolically, and nitrogen oxides, generated under pathological conditions, interact with proteins, unsaturated fatty acids, and thiol-containing small molecules like glutathione (GSH) to create nitroso (RNO) or nitro (RNO_2) compounds [2, 3]. Nitrogen dioxide radicals ($NO_2\cdot$) interact with unsaturated fatty acids, generating electrophilic bioactive lipids. These lipids interact with nucleophilic amino acid sites, primarily cysteines, in transcriptional regulatory proteins or enzymes associated with metabolism, cell signaling, and redox homeostasis, resulting in the formation of covalent Michael addition adducts [4]. Nitrated fatty acids perform various functions via post-translational modification (PTM) [5].

The products of the nitration reaction of unsaturated fatty acids are dictated by their structural characteristics, with product formation significantly influenced by the distinct chemical reactivities of conjugated and unconjugated alkenes present in these fatty acids. Two distinct response pathways have been suggested for the NO_2-FAs [6]. The first mechanism involves the dehydrogenation of the bis-allylic carbon, resulting in the formation of alkyl radicals through free radicals (OH·, OOH·, $NO_2\cdot$). This is a typical response in the generation of alternative non-enzymatic lipid oxidation products. Following dehydrogenation, the fatty acid experiences double bond rearrangement and the incorporation of molecular oxygen (O_2), which reacts with peroxyl or nitro radicals to provide nitroolefin derivatives [7]. The second method involves the direct addition of $NO_2\cdot$ to produce a carbon-centered radical. This radical can undergo further oxidation, with or without $NO_2\cdot$, to yield electrophilic nitroolefin compounds. The reaction with an additional molecule of $NO_2\cdot$ might yield unstable nitro-nitroso or dinitro compounds, which swiftly disintegrate to produce electrophilic nitroolefin products, accompanied by the release of HNO_2 [8]. The nitration of unsaturated fatty acids is contingent upon the ambient O_2 concentration. Nitrification predominates at low O_2 concentrations, while lipid peroxidation occurs at elevated O_2 values. Despite the presence of several NO_2-FAs, only nitro-linoleic acid is significantly prevalent. Nitro-linoleic acid (NO_2-LA) is the primary nitrated fatty acid produced in mouse research and human clinical trials.

7.1.2 Signal Regulation by Nitro-Fatty Acids

Nitrated fatty acids play a crucial role in regulating various signaling pathways linked to anti-inflammatory effects. Their electrophilic properties are leveraged to modify the overall structure of proteins by facilitating the formation of new chemical bonds, consequently changing their function. Structural changes in proteins are referred to as post-translational modifications (PTM). Post-translational modifications of proteins and transcription factors influence gene expression [9].

7.1.2.1 Initiation of Nrf2/Keap-1 Pathway

The transcription factor Nrf2 is a reactant to oxidative stress that regulates the expression of genes associated with cellular antioxidants and protective roles. Nrf2 is present at low levels in the cytoplasm, where it is associated with the inhibitor Keap1 (Kelch-like ECH-associated protein 1) to sustain cellular homeostasis. The nitro-fatty acid pro-electric potential site activates Nrf2 and enhances antioxidant response element (ARE), two of which form a dimer in order for NF-κB to translocate to the nucleus C–S bond, thereby increasing the levels of antioxidants such as GSH [10–12].

7.1.2.2 Inhibition of NF-κB Pathway

NF-κB (nuclear factor-κB) governs cellular survival and proliferation and facilitates inflammatory responses [13]. It is crucial in the body's immunological response and the eradication of localized infections. NF-κB comprises five distinct subunits (p50, p65, p52, RelB, and c-Rel), two of which dimerize to facilitate NF-κB's translocation to the nucleus, where it binds to DNA and enhances the transcription of inflammation-associated genes. In the cytoplasm, the NF-κB inhibitory protein (IκB) interacts with NF-κB, therefore obstructing its translocation into the nucleus. Upon receiving an inflammatory stimulus, IκB kinase (IKK) phosphorylates IκB, leading to its degradation and enabling NF-κB to activate and perform its functions. If this route is obstructed, the inflammatory response is suppressed.

Nitro-fatty acids impede NF-κB's DNA binding through the nitro-alkylation of cysteine residues in the P65 subunit [14]. After endotoxin stimulation of macrophages, nitro-fatty acids reduce the expression of IL-6, tumor necrosis factor (TNF-α), and monocyte chemotactic protein-1 (MCP-1), thus inhibiting the inflammatory response. Nitro-fatty acids also inhibit monocyte-endothelial cell interactions by decreasing the expression of vascular cell adhesion molecule-1 (VCAM-1), which is regulated by NF-κB.

7.1.2.3 Promotion of PPARγ Pathway

Peroxisome proliferator-activated receptor γ (PPARγ) is a nuclear receptor predominantly found in monocyte macrophages, adipocytes, smooth muscle cells, and endothelial cells. Its functions are intricately associated with the response to inflammatory stimuli, the regulation of glucose metabolism, and the differentiation of adipocytes. PPARγ regulates fat distribution in peripheral organs and decreases adipokine levels, thereby enhancing insulin sensitivity. Furthermore, PPARγ facilitates lipid scavenging in monocyte macrophages and suppresses the expression of inflammatory mediators, including interferon γ and inducible nitric oxide synthase (iNOS or NOS2) [15].

Nitrated fatty acids serve as significant ligands for PPARγ. Nitrated fatty acids exhibit specific binding to arginine and glutamate residues of PPARγ [16]. Activation of PPARγ enhances the expression of CD36 in macrophages and facilitates the differentiation of adipocytes [17]. Nitro-fatty acids up-regulate the genes encoding GLUT4, CBP-1, and PGC-1α [18], which enhance insulin sensitivity and decrease adiposity accumulation and inflammatory factors. In contrast to pharmacological PPARγ agonists like the thiazolidinedione-class insulin sensitizer rosiglitazone, nitro-fatty acids function solely as partial agonists [19], thereby minimizing the risk of peripheral edema, weight gain, and negative cardiovascular outcomes.

7.1.2.4 Induction of Heat Shock Response

The heat shock response (HSR) is regulated by proteins and transcription factors, facilitating the expression of anti-inflammatory genes and genes that confer cell protection [20]. Research indicates that HSR is a significant signaling pathway activated by nitrated fatty acids. The stress response is initiated by the induction of heat shock proteins (HSP), which reduces the accumulation of unfolded proteins and mitigates the endoplasmic reticulum stress response. Nitro-fatty acids influence human umbilical vein endothelial cells by enhancing the expression of target genes governed by the Nrf2 pathway, along with the heat shock factors HSF-1 and HSF-2. The expression of HSP70 is associated with the dosage [20]. It plays a significant role in diminishing inflammatory responses and mitigating cellular stress.

7.1.3 The Physiological and Therapeutic Effects of Nitro-Fatty Acids

7.1.3.1 Diabetes Mellitus and Metabolic Syndrome

The rising prevalence of obesity throughout the population elevates the risk of endocrine and metabolic disorders. Thiazolidinediones, specifically rosiglitazone and pioglitazone, which function as PPARγ agonists, elevate the risk of cardiovascular and oncological diseases. Hence, the utilization of these medications for type 2 diabetes management is declining [21]. Nitro-fatty acids possessing electrophilic characteristics can effectively mitigate impaired glucose tolerance and insulin resistance. Keto ester derivatives of docosahexaenoic acid (DHA) were shown to lower blood glucose levels without adverse effects in leptin receptor-deficient mouse models and obese rat models [21]. Moreover, the injection of eicosapentaenoic acid (EPA) and DHA enhanced vascular function in patients with type 2 diabetes relative to the placebo group. Nitrated fatty acids can diminish oxide synthesis, therefore enhancing the sensitivity of vascular smooth muscle cells to NO-mediated vasodilation [22]. Resolvin D1 (RvD1), produced from omega-3 fatty acids, enhances insulin sensitivity by mitigating chronic inflammation linked to obesity (reducing IL-6 production) while concurrently elevating lipocalin levels. Arachidonic acid (AA) metabolites, including heparin, function as insulin sensitizers and stimulate insulin secretion from pancreatic islet cells. Nitro-fatty acids have demonstrated efficacy in mitigating metabolic dysfunction in obese animals. Treatment for leptin receptor-deficient mice and those subjected to a high-fat diet with nitro-oleic acid (NO_2-OA) diminished glucose tolerance and exacerbated insulin resistance. While rosiglitazone operates via the PPARγ pathway, nitroxide does not exhibit the adverse consequence of weight gain. The administration of nitro-fatty acids to obese Zucker rats significantly decreased body weight, accompanied by reductions in plasma triglycerides and lipid peroxidation products, as well as an increase in anti-inflammatory high-density lipoproteins [23], ultimately alleviating insulin resistance in the organism.

7.1.3.2 Renal Disease

Nephritis and renal failure typically arise from chronic inflammatory conditions such as diabetes and hypertension. In a mouse model of adriamycin-induced focal glomerulosclerosis, NO_2-OA pretreatment diminished creatinine levels, mitigated glomerulosclerosis, tubulointerstitial fibrosis, renal inflammation, and decreased urinary lipid peroxidation products [24].

Following nitro-fatty acid treatment in a multiorgan endotoxemia model, the concentrations of inflammatory mediators, including tumor necrosis factor-α (TNF-α), MCP-1, intercellular adhesion molecule-1 (ICAM-1), VCAM-1, and prostaglandin E2 (PGE2), were reduced. Renal protection was observed in mice administered nitro-linoleic acid following 30 min of ischemic reperfusion of bilateral renal arteries [25]. Following 14 days of NO_2-OA treatment, obese Zucker rats exhibited a reduction in food consumption and a gain in body weight, alongside a notable decline in plasma triglycerides, restoration of plasma free fatty acid levels, and an elevation in plasma high-density lipoproteins. Treatment with NO_2-OA dramatically reduced proteinuria in Zucker rats, indicating that nitro-fatty acids provide a protective effect against diabetes mellitus [23]. Nrf2 inducers protected renal cells from oxidative stress [26]. This indicates that nitrified fatty acids safeguard and preserve kidney physiological function by activating the Nrf2/Keap-1 pathway and diminishing NF-κB-dependent gene expression.

7.1.3.3 Cardiovascular Disease

The detrimental effects of activated platelets, monocytes, and neutrophils can contribute to atherosclerosis and hence vascular blockage. Subsequent occurrences include progressive plaque deposition, myocardial infarction, peripheral vascular disease, and stroke. Following intravenous administration of NO_2-OA135, vascular inflammation was diminished through pathways including leukocyte adhesion, endothelial NF-κB activation, and TLR4 signaling activation. Following NO_2-LA administration, thrombin-induced platelet aggregation was reduced by a mechanism linked to increased cAMP levels. Furthermore, NO_2-LA diminished platelet P-selectin expression in an NO-independent manner, resulting in decreased platelet adhesion and improved anti-aggregation [27]. Nitro arachidonic acid (NO_2-AA) suppresses platelet aggregation by suppressing the production of thromboxane B2 (TXB2) via prostaglandin endoperoxide synthase (PGHS) [28].

Moreover, leukocyte aggregation and unstable plaque in atherosclerotic lesions originate from the activation of neutrophils and platelets [29]. Significantly, macrophages are attracted to the plaque, where reactive oxygen species oxidize LDLs, prompting the transformation of macrophages into atherogenic foam cells. NO_2-OA therapy diminishes lesion development in the mouse model of atherosclerosis by decreasing inflammatory cell accumulation and chemokine synthesis. NO_2-OA inhibits low-density lipoprotein (LDL)-induced phosphorylation of STAT1, hence hindering lipid absorption and foam cell production in macrophages [30].

7.2 The Role of Nitrate in the Prevention and Treatment of Hypertension

Jianyu Gu

Hypertension is defined as an office-based systolic blood pressure $\geq$ 140 mmHg and/or a diastolic blood pressure $\geq$ 90 mmHg without the use of antihypertensive drugs [31]. Hypertension is an important risk factor for a wide range of cardiovascular risk events, and lowering blood pressure is the most important principle and modality of benefit in the treatment [31, 32]. A diet rich in fresh vegetables and fruits is beneficial to blood pressure control in people with hypertension, as the guidelines suggested [31]. Nitrates in green leafy vegetables show positive effects on blood pressure control, which is an important new chemical basis for the health-promoting effects of green vegetables.

Nitrates or nitrate-rich foods have shown beneficial physiological effects in hypertensive models. Diastolic blood pressure and mean arterial pressure were reduced in 17 healthy subjects after direct administration of inorganic nitrates [33]. Consumption of nitrate-enriched beetroot juice by healthy subjects also revealed significant reduction in blood pressure after 3 h [34]. After receiving 250 mL of nitrate-rich beetroot juice treatment, the clinic-measured blood pressure, ambulatory blood pressure and home blood pressure of 68 hypertensive patients decreased significantly [35]. This suggests that dietary nitrate supplementation can safely and effectively lower blood pressure and has preventive and therapeutic effects on hypertension. However, healthy volunteers or hypertensive patients can see a significant increase in blood pressure after continuous use of antibacterial mouthwash [36]. This indicates that nitrates can be reduced to nitrites and further reduced to NO under the action of oral flora, which plays an important physiological role in maintaining blood pressure.

7.3 Protective Effect of Nitrate Against Myocardial Ischemia-Reperfusion Injury

Xiaoyu Li

Acute myocardial infarction (AMI) can lead to severe adverse cardiovascular events, including malignant arrhythmias, cardiogenic shock, and sudden death, and represents a significant threat to human health. Reperfusion therapy, including thrombolysis and percutaneous coronary intervention, can occlude obstructed coronary arteries, reinstate blood flow to cardiac tissues, preserve ischemic myocardial tissues, diminish the extent of myocardial infarction, and enhance patient prognosis. Consequently, reperfusion treatment has emerged as the most efficacious method for managing AMI. Nonetheless, reperfusion therapy may inflict harm on heart tissues, referred to as myocardial ischemia-reperfusion injury (MIRI) [37]. MIRI primarily presents as arrhythmias such as ventricular pre-systole and ventricular tachycardia, along with persistent cardiac dysfunction, microvascular blockage, and programmed cardiac myocyte death [38]. Therefore, the pursuit of efficient targets and strategies to safeguard against ischemia-reperfusion injury is presently a prominent research focus in myocardial protection.

Several animal studies have revealed that nitrates and nitrites reduce infarct size and safeguard neurological and cardiac function in models of myocardial ischemia-reperfusion injury [39–42]. The half-life of nitrates (5–6 h) significantly exceeds that of nitrites (20–30 min), allowing nitrates to serve as a precursor medication to enhance nitrite levels and extend chronotropic effects in blood and tissues. The nitrate-nitrite-NO pathway generates not only NO but also other reactive nitrogen species (RNS) that can nitrosylate (-SNO) and nitroxylate ($-NO_2$) proteins, thus modifying their functionality. The cytoprotective effect may be mechanistically ascribed to the S-nitrosylation of complex I in the mitochondrial respiratory chain, resulting in diminished superoxide formation, including reactive oxygen species (ROS), or to a decrease in apoptosis due to the inhibition of mitochondrial permeability transition pore opening. Furthermore, the cytoprotective effect may also play a role in the activation of ATP-sensitive potassium channels (KATP) [43, 44]. Alongside these mitochondrial effects, vasodilation of coronary arteries, suppression of platelet aggregation, and inhibition of leukocyte adhesion may also play a role in the protective effects of nitrates, nitrites, and NO. [43, 44] Additional potential advantageous effects of nitrates and nitrites in the pathophysiology of myocardial ischemia-reperfusion injury encompass vasodilation [45], suppression of platelet aggregation [34], enhanced resistance to leukocyte adhesion [46], amelioration of endothelial function [47], and inhibition of nicotinamide adenine dinucleotide phosphate oxidase (NADPH) production (Fig. 7.2) [48].

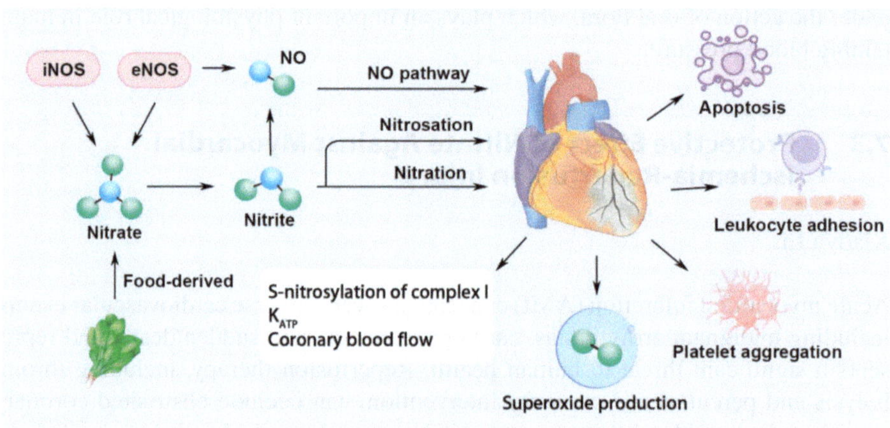

Fig. 7.2 The protective role of nitrates and nitrites in myocardial ischemia-reperfusion injury

Two clinical studies explored the effects of intravenous or intracoronary nitrite injection in patients with acute myocardial infarction undergoing percutaneous coronary intervention. Although intravenous nitrite lacks a protective effect in patients with acute myocardial infarction [49], intracoronary nitrite injection may positively affect acute arrhythmia and major adverse cardiac events at 1 year [50]. Furthermore, intravenous nitrite administration during resuscitation from out-of-hospital cardiac arrest did not improve survival in hospitalized patients [51]. In a recent study, although nitrate therapy did not have any protective effect on troponin T release or other biomarkers of cardiac injury in subjects after coronary artery bypass grafting (CABG), perioperative bleeding was significantly reduced in nitrate-treated subjects [52]. In conclusion, nitrates exert a protective effect against myocardial ischemia-reperfusion injury.

7.4 Protective Effects of Nitrates Against Other Cardiovascular Diseases

Xiaoyu Li

In December 2020, the *Journal of the American College of Cardiology* (JACC) released an authoritative report on the worldwide prevalence of cardiovascular disease and its risk factors. This research evaluated the health of populations in 204 nations from 1990 to 2019. The findings indicated that the prevalence of cardiovascular illness escalated from 271 to 523 million, reflecting a growth rate of 92.99%. Concurrently, fatalities attributed to cardiovascular disease rose from 12.1 to 18.6 million, establishing it as the condition with the greatest fatality rate. Consequently, the efficient prevention of cardiovascular illnesses has become a significant scientific concern.

Nitrates can produce NO-like effects and improve cardiovascular function in conditions of NO deficiency [53]. Dietary nitrate application has demonstrated positive effects on cardiac health in various animal models. In a model of adriamycin-induced cardiac insufficiency in mice, the administration of nitrates enhanced left ventricular systolic pressure, end-diastolic pressure, and ejection fraction [54]. Nitrate-enriched beet juice enhanced resting blood flow and vascular conductance in rats with heart failure, exhibiting more significant improvements during exercise [55]. Dietary nitrates demonstrated positive effects on exercise performance in patients with heart failure [56]. Nitrate intake preserved ejection fraction during exercise and improved exercise tolerance in elderly patients with heart failure [57]. Therefore, nitrates are anticipated to enhance the quality of life in patients with heart failure.

Furthermore, it has been shown that NO inhibits platelet aggregation and adhesion, as well as leukocyte aggregation, thus decreasing the recruitment and formation of platelet-leukocyte aggregates [58]. Nitrites and nitrates preconditioning reduced platelet aggregation and extended bleeding time, while a low nitrate diet for mice produced contrary effects [59]. For patients with endothelial dysfunction and mild hypercholesterolemia, dietary nitrates suppressed platelet P-selectin expression and platelet-leukocyte aggregates, thereby contributing to a reduction in thrombosis [60]. In conclusion, nitrates serve a vital protective role in various cardiovascular diseases.

Compliance with Ethical Standards This chapter is a review of previously published accounts, as such, no animal or human studies were performed.

References

1. O'Donnell VB, Eiserich JP, Chumley PH, et al. Nitration of unsaturated fatty acids by NO-derived reactive nitrogen species peroxynitrite, nitrous acid, nitrogen dioxide, and nitronium ion. Chem Res Toxicol. 1999;12(1):83–92.
2. Bryan NS. Application of nitric oxide in drug discovery and development. Expert Opin Drug Discov. 2011;6(11):1139–54.
3. Weitzberg E, Lundberg JO. Novel aspects of dietary nitrate and human health. Annu Rev Nutr. 2013;33:129–59.
4. Schopfer FJ, Cipollina C, Freeman BA. Formation and signaling actions of electrophilic lipids. Chem Rev. 2011;111(10):5997–6021.
5. Turell L, Vitturi DA, Coitiño EL, et al. The chemical basis of thiol addition to nitro-conjugated linoleic acid, a protective cell-signaling lipid. J Biol Chem. 2017;292(4):1145–59.
6. Woodcock SR, Bonacci G, Gelhaus SL, et al. Nitrated fatty acids: synthesis and measurement. Free Radic Biol Med. 2013;59:14–26.
7. Yin HY, Xu LB, Porter NA. Free radical lipid peroxidation: mechanisms and analysis. Chem Rev. 2011;111(10):5944–72.
8. Pryor WA, Lightsey JW, Church DF. Reaction of nitrogen dioxide with alkenes and polyunsaturated fatty acids: addition and hydrogen-abstraction mechanisms. J Am Chem Soc. 1982;104(24):6685–92.
9. Jones AM, Thompson C, Wylie LJ, et al. Dietary nitrate and physical performance. Annu Rev Nutr. 2018;38:303–28.

10. Batthyany C, Schopfer FJ, Baker PR, et al. Reversible post-translational modification of proteins by nitrated fatty acids *in vivo*. J Biol Chem. 2006;281(29):20450–63.
11. Tsujita T, Li L, Nakajima H, et al. Nitro-fatty acids and cyclopentenone prostaglandins share strategies to activate the Keap1-Nrf2 system: a study using green fluorescent protein transgenic zebrafish. Genes Cells. 2011;16(1):46–57.
12. Kansanen E, Jyrkkänen HK, Volger OL, et al. Nrf2-dependent and -independent responses to nitro-fatty acids in human endothelial cells: identification of heat shock response as the major pathway activated by nitro-oleic acid. J Biol Chem. 2009;284(48):33233–41.
13. Tak PP, Firestein GS. NF-κB: a key role in inflammatory diseases. J Clin Invest. 2001;107(1):7–11.
14. Cui T, Schopfer FJ, Zhang J, et al. Nitrated fatty acids: endogenous anti-inflammatory signaling mediators. J Biol Chem. 2006;281(47):35686–98.
15. Tontonoz P, Spiegelman BM. Fat and beyond: the diverse biology of Ppargamma. Annu Rev Biochem. 2008;77:289–312.
16. Li Y, Zhang JF, Schopfer FJ, et al. Molecular recognition of nitrated fatty acids by PPARgamma. Nat Struct Mol Biol. 2008;15(8):865–7.
17. Schopfer FJ, Lin YM, Baker PR, et al. Nitrolinoleic acid: an endogenous peroxisome proliferator-activated receptor gamma ligand. Proc Natl Acad Sci U S A. 2005;102(7):2340–5.
18. Groeger AL, Cipollina C, Cole MP, et al. Cyclooxygenase-2 generates anti-inflammatory mediators from omega-3 fatty acids. Nat Chem Biol. 2010;6(6):433–41.
19. Baker PR, Lin Y, Schopfer FJ, et al. Fatty acid transduction of nitric oxide signaling: multiple nitrated unsaturated fatty acid derivatives exist in human blood and urine and serve as endogenous peroxisome proliferator-activated receptor ligands. J Biol Chem. 2005;280(51):42464–75.
20. Woodcock J, Sharfstein JM, Hamburg M. Regulatory action on rosiglitazone by the U.S. Food and Drug Administration. N Engl J Med. 2010;363(16):1489–91.
21. Yamamoto K, Itoh T, Abe D, et al. Identification of putative metabolites of docosahexaenoic acid as potent PPARgamma agonists and antidiabetic agents. Bioorg Med Chem Lett. 2005;15(3):517–22.
22. Stirban A, Nandrean S, Gotting C, et al. Effects of n-3 fatty acids on macro-and microvascular function in subjects with type 2 diabetes mellitus. Am J Clin Nutr. 2010;91(3):808–13.
23. Wang HP, Liu HY, Jia ZJ, et al. Effects of endogenous PPAR agonist nitro-oleic acid on metabolic syndrome in obese Zucker rats. PPAR Res. 2010;2010:601562.
24. Liu SS, Jia ZJ, Zhou L, et al. Nitro-oleic acid protects against adriamycin-induced nephropathy in mice. Am J Physiol Ren Physiol. 2013;305(11):F11533–41.
25. Liu HY, Jia ZJ, Soodvilai S, et al. Nitro-oleic acid protects the mouse kidney from ischemia and reperfusion injury. Am J Physiol Ren Physiol. 2008;295(4):F942–9.
26. Zhu H, Zhang L, Amin AR, et al. Coordinated upregulation of a series of endogenous antioxidants and phase 2 enzymes as a novel strategy for protecting renal tubular cells from oxidative and electrophilic stress. Exp Biol Med (Maywood). 2008;233(6):753–65.
27. Coles B, Bloodsworth A, Eiserich JP, et al. Nitrolinoleate inhibits platelet activation by attenuating calcium mobilization and inducing phosphorylation of vasodilator-stimulated phosphoprotein through elevation of camp. J Biol Chem. 2002;277(8):5832–40.
28. Trostchansky A, Bonilla L, Thomas CP, et al. Nitroarachidonic acid, a novel peroxidase inhibitor of prostaglandin endoperoxide H synthases 1 and 2. J Biol Chem. 2011;286(15):12891–900.
29. Della Bona R, Cardillo MT, Leo M, et al. Polymorphonuclear neutrophils and instability of the atherosclerotic plaque: a causative role? Inflamm Res. 2013;62(6):537–50.
30. Rudolph TK, Rudolph V, Edreira MM, et al. Nitro-fatty acids reduce atherosclerosis in apolipoprotein E-deficient mice. Arterioscler Thromb Vasc Biol. 2010;30(5):938–45.
31. China Hypertension Prevention and Control Guidelines Revision Committee, Hypertension Alliance (China), Chinese Society of Cardiovascular Diseases of the Chinese Medical Association Chinese Physicians Association Hypertension Professional Committee, et al. Chinese guidelines for the prevention and treatment of hypertension (2018 revision). Chin Cardiovas J. 2019;24(1):24–56.

32. Chobanian AV, Bakris GL, Black HR, et al. The seventh report of the joint national committee on prevention, detection, evaluation, and treatment of high blood pressure: the JNC 7 report. JAMA. 2003;289(19):2560–72.
33. Larsen FJ, Ekblom B, Sahlin K, et al. Effects of dietary nitrate on blood pressure in healthy volunteers. N Engl J Med. 2006;355(26):2792–3.
34. Webb AJ, Patel N, Loukogeorgakis S, et al. Acute blood pressure lowering, vasoprotective, and antiplatelet properties of dietary nitrate via bioconversion to nitrite. Hypertension. 2008;51(3):784–90.
35. Kapil V, Khambata RS, Robertson A, et al. Dietary nitrate provides sustained blood pressure lowering in hypertensive patients:a randomized, phase 2, double-blind, placebo-controlled study. Hypertension. 2015;65(2):320–7.
36. Bondonno CP, Liu AH, Croft KD, et al. Antibacterial mouthwash blunts oral nitrate reduction and increases blood pressure in treated hypertensive men and women. Am J Hypertens. 2015;28(5):572–5.
37. Hausenloy DJ, Yellon DM. Myocardial ischemia-reperfusion injury: a neglected therapeutic target. J Clin Invest. 2013;123(1):92–100.
38. Vander Heide RS, Steenbergen C. Cardioprotection and myocardial reperfusion: pitfalls to clinical application. Circ Res. 2013;113(4):464–77.
39. Webb A, Bond R, Mclean P, et al. Reduction of nitrite to nitric oxide during ischemia protects against myocardial ischemia-reperfusion damage. Proc Natl Acad Sci USA. 2004;101(37):13683–8.
40. Duranski MR, Greer JJ, Dejam A, et al. Cytoprotective effects of nitrite during *in vivo* ischemia-reperfusion of the heart and liver. J Clin Invest. 2005;115(5):1232–40.
41. Omar SA, Webb AJ, Lundberg JO, et al. Therapeutic effects of inorganic nitrate and nitrite in cardiovascular and metabolic diseases. J Intern Med. 2016;279(4):315–36.
42. Dezfulian C, Raat N, Shiva S, et al. Role of the anion nitrite in ischemia-reperfusion cytoprotection and therapeutics. Cardiovasc Res. 2007;75(2):327–38.
43. Shiva S, Sack MN, Greer JJ, et al. Nitrite augments tolerance to ischemia/reperfusion injury via the modulation of mitochondrial electron transfer. J Exp Med. 2007;204(9):2089–102.
44. Chouchani ET, Methner C, Nadtochiy SM, et al. Cardioprotection by S-nitrosation of a cysteine switch on mitochondrial complex I. Nat Med. 2013;19(6):753–9.
45. Modin A, Björne H, Herulf M, et al. Nitrite-derived nitric oxide: a possible mediator of 'acidic-metabolic' vasodilation. Acta Physiol Scand. 2001;171(1):9–16.
46. Jädert C, Petersson J, Massena S, et al. Decreased leukocyte recruitment by inorganic nitrate and nitrite in microvascular inflammation and NSAID-induced intestinal injury. Free Radic Biol Med. 2012;52(3):683–92.
47. Rodriguez-Mateos A, Hezel M, Aydin H, et al. Interactions between cocoa flavanols and inorganic nitrate: additive effects on endothelial function at achievable dietary amounts. Free Radic Biol Med. 2015;80:121–8.
48. Cordero-Herrera I, Kozyra M, Zhuge Z, et al. Amp-activated protein kinase activation and NADPH oxidase inhibition by inorganic nitrate and nitrite prevent liver steatosis. Proc Natl Acad Sci USA. 2019;116(1):217–26.
49. Siddiqi N, Neil C, Bruce M, et al. Intravenous sodium nitrite in acute ST-elevation myocardial infarction: a randomized controlled trial (NIAMI). Eur Heart J. 2014;35(19):1255–62.
50. Jones DA, Pellaton C, Velmurugan S, et al. Randomized phase 2 trial of intracoronary nitrite during acute myocardial infarction. Circ Res. 2015;116(3):437–47.
51. Kim F, Maynard C, Dezfulian C, et al. Effect of out-of-hospital sodium nitrite on survival to hospital admission after cardiac arrest: a randomized clinical trial. JAMA. 2021;325(2):138–45.
52. Eriksson KE, Eidhagen F, Liska J, et al. Effects of inorganic nitrate on ischaemia-reperfusion injury after coronary artery bypass surgery: a randomised controlled trial. Br J Anaesth. 2021;127(4):547–55.
53. Paulus WJ, Tschöpe C. A novel paradigm for heart failure with preserved ejection fraction: comorbidities drive myocardial dysfunction and remodeling through coronary microvascular endothelial inflammation. J Am Coll Cardiol. 2013;62(4):263–71.

54. Zhu SG, Kukreja RC, Das A, et al. Dietary nitrate supplementation protects against doxorubicin-induced cardiomyopathy by improving mitochondrial function. J Am Coll Cardiol. 2011;57(21):2181–9.
55. Ferguson SK, Hirai DM, Copp SW, et al. Effects of nitrate supplementation via beetroot juice on contracting rat skeletal muscle microvascular oxygen pressure dynamics. Respir Physiol Neurobiol. 2013;187(3):250–5.
56. Coggan AR, Leibowitz JL, Spearie CA, et al. Acute dietary nitrate intake improves muscle contractile function in patients with heart failure: a double-blind, placebo-controlled, randomized trial. Circ Heart Fail. 2015;8(5):914–20.
57. Zamani P, Rawat D, Shiva-Kumar P, et al. Effect of inorganic nitrate on exercise capacity in heart failure with preserved ejection fraction. Circulation. 2015;131(4):371–80.
58. Chung AW, Radomski A, Alonso-Escolano D, et al. Platelet-leukocyte aggregation induced by PAR agonists: regulation by nitric oxide and matrix metalloproteinases. Br J Pharmacol. 2004;143(7):845–55.
59. Park JW, Piknova B, Huang PL, et al. Effect of blood nitrite and nitrate levels on murine platelet function. PLoS One. 2013;8(2):e55699.
60. Velmurugan S, Gan JM, Rathod KS, et al. Dietary nitrate improves vascular function in patients with hypercholesterolemia: a randomized, double-blind, placebo-controlled study. Am J Clin Nutr. 2016;103(1):25–38.

Tumor Prevention, Therapy, and Radiation Protection

8

Yuanyong Feng, Haibo Wang, Lei Hu, and Xiaoyu Feng

The nitrate-nitrite cycle is essential for regulating the physiological levels of NO in blood and tissues, influencing numerous biological functions. Research conducted by Academician Songlin Wang's group has demonstrated the various advantageous roles of nitrates, including their effectiveness in preventing systemic damage from whole-body gamma-ray radiotherapy, alleviating liver damage and degeneration, reducing insufficient salivation, and improving chemosensitivity.

8.1 Nitrate Enhances Cisplatin Chemosensitivity

Yuanyong Feng

The 5-year survival rate for oral squamous cell carcinoma (OSCC) has shown minimal variation over recent decades, primarily due to the failure of chemotherapy [1–3]. Cisplatin-based chemotherapy serves as the primary treatment for patients with advanced OSCC, particularly in cases of recurrent or metastatic OSCC [4]. Unfortunately, over 30% of OSCC patients demonstrate resistance to cisplatin, with additional patients developing resistance following multiple chemotherapy cycles [5]. Hybrid therapies utilizing platinum compounds alongside other chemotherapeutic agents for the treatment of OSCC frequently result in significant and

Y. Feng
The Affiliated Hospital of Qingdao University, Qingdao, China

H. Wang
Department of Biochemistry and Molecular Biology, School of Basic Medical Sciences, Capital Medical University, Beijing, China

L. Hu (✉) · X. Feng
Beijing Stomatological Hospital, Capital Medical University, Beijing, China
e-mail: hulei@ccmu.edu.cn

S. Wang (ed.), *Inorganic Nitrate and Human Body Homeostasis*,
https://doi.org/10.1007/978-981-95-2824-0_8

unpredictable adverse effects. These include hepatorenal toxicity, gastrointestinal and hematological toxicity, as well as compromised immune cell function. The side effects notably diminish the efficacy of platinum-based chemotherapy, leading to chemotherapy failure [6]. Consequently, formulating strategies to improve chemotherapy sensitivity while reducing adverse effects is a significant challenge in cancer treatment.

8.1.1 NO and Chemotherapeutic Sensitization

Nitrate is a crucial precursor for the synthesis of NO in the human body [7]. Synthetic nitrates function as NO donors, delivering a sustained and elevated concentration of NO, which may serve as a potential anti-tumor agent. Their effectiveness, however, depends on the tumor microenvironment [8]. For example, the combination of doxorubicin and nitroglycerin has demonstrated a reduction in tumor growth at low doses (0.1 nmol/L) [9], whereas sodium nitroprusside (1 mmol/L) can induce apoptosis in T-cell lymphoma cells [10]. NO donors, including DETA/NO, have been shown to enhance cisplatin-mediated cytotoxicity in head and neck squamous cell carcinoma [11]. In xenografted melanoma models, the combination of cisplatin (1 mg/kg) and 10 μmol/L DETA/NO leads to a reduction in tumor volume in nude mice [12]. The chemosensitization associated with these NO donors is related to alterations in plasma nitrate/nitrite levels and the activation of protein kinase B (AKT). Increased nitrate/nitrite levels in rats on a high-fat diet result in a decrease in phosphorylated AKT (pAKT), indicating that nitrates may influence the pAKT signaling pathway [13].

The phosphorylation of AKT is intricately associated with multiple gene regulatory processes, with the developmental and DNA damage response regulatory gene (*REDD1*) being a significant contributor. *REDD1*, a gene associated with DNA repair, is essential for safeguarding specific cell types against DNA damage induced by cisplatin [14]. *REDD1*, identified in 2002, functions as an upstream repressor essential for inhibiting mammalian target of rapamycin (mTOR) activity and serves as a key regulator of mTOR signaling [15]. This gene has been identified as a stress-response gene, activated by hypoxia, oxidative stress [16], and various stress conditions, including glucocorticoid therapy and chemotherapy [17, 18]. *REDD1*'s response to these stressors increases AKT phosphorylation in *REDD1*-expressing cells, resulting in strengthened resistance to chemotherapeutic agents. AKT activation appears to play a role in promoting cell survival during stress conditions [19–21]. The repair of DNA damage induced by cisplatin is frequently associated with resistance to chemotherapy. Our preliminary study suggests that nitrates may enhance the chemosensitivity of cisplatin in OSCC cells. This effect is mediated by a decrease in *REDD1* expression and the activation of AKT signaling, which improves the therapeutic response to cisplatin.

8.1.2 Chemotherapy Sensitization of Nitrates in OSCC: In Vivo and In Vitro

We investigated the mechanism by which *REDD1* modulates OSCC sensitization to cisplatin chemotherapy in the presence of nitrates through both in vitro and in vivo experiments. In addition, we examined the relationship between nitrate-induced cisplatin chemosensitization and the *REDD1*/AKT signaling pathway in the context of oxygen-serum deprivation (OSD) conditions.

8.1.2.1 Impact of Nitrates on Cisplatin Chemotherapy in Mouse Xenograft Tumors

To examine the in vivo impact of nitrates on the chemosensitization of OSCC to cisplatin, xenograft tumor models were created by subcutaneously injecting CAL27 cells and SCC VII cells into the axillary region of nude mice and C3H mice, respectively. Drinking water with nitrates was freshly prepared, and cisplatin was administered through intraperitoneal injection. The findings indicated that nitrate treatment significantly improved the inhibitory effect of cisplatin on tumor growth (Fig. 8.1).

8.1.2.2 The Combined Inhibitory Effect of Nitrate and Cisplatin on OSCC Proliferation

CAL27 and SCC VII cell lines were cultured under oxygen-serum deprivation (OSD) conditions to simulate the hypoxic tumor environment in vivo. In vitro studies demonstrated that the combination of cisplatin and nitrate produced a more pronounced inhibitory effect on OSCC cell proliferation than cisplatin alone. Nitrate augmented the cytotoxic effects of cisplatin, leading to heightened apoptosis in OSCC cells. Furthermore, nitrate enhanced cisplatin-induced cell cycle arrest in the G2/M phase.

The regulation of *REDD1* is significant in this context, as decreased *REDD1* levels increased the sensitivity of OSCC cells to apoptosis, whereas elevated *REDD1* levels provided resistance to apoptotic stimuli [22]. The combination of cisplatin

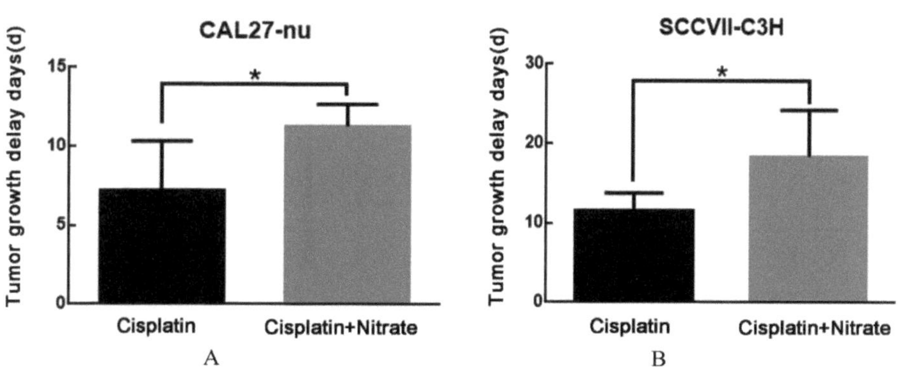

Fig. 8.1 Combined treatment with nitrate and cisplatin inhibited the proliferation of OSCC transplanted tumors (*$P < 0.05$). (**a**) C3H mice. (**b**) Nude mice

and nitrate resulted in a higher apoptosis rate under OSD conditions compared to cisplatin treatment alone. Treatment with both cisplatin and nitrate resulted in an increased number of cells arrested in the G2/M phase, indicating that nitrate enhances the chemosensitivity of OSCC cells under OSD conditions (Fig. 8.2).

8.1.2.3 Chemotherapy Sensitization of Nitrates in OSCC Cell Lines Through the REDD1/AKT Signaling Pathway

Prior research indicated that reduced *REDD1* expression correlates with heightened sensitivity to cisplatin in cancer cells, given that *REDD1* is crucial in preventing cisplatin-induced DNA damage [14]. This study examined the role of the *REDD1*/AKT signaling pathway in the sensitization to cisplatin chemotherapy enhanced by nitrates. The findings suggested that nitrate-induced suppression of *REDD1* expression may enhance the sensitivity of OSCC cells to cisplatin therapy. Specifically, the diminished expression of *REDD1* was associated with a reduction in pAKT levels, which is known to enhance cisplatin chemosensitivity.

Blocking AKT signaling is known to enhance the anti-tumor efficacy of chemotherapeutic agents [15, 23, 24]. Additionally, inhibiting AKT phosphorylation in *REDD1*-expressing cancer cells can augment their sensitivity to these agents [14]. Therefore, agents that inhibit both *REDD1* and pAKT may enhance the susceptibility of cancer cells to cisplatin. To assess the role of nitrate in enhancing cisplatin inhibition in OSCC cells through the *REDD1*/AKT pathway, we transfected cells with *REDD1* siRNA. The knockdown of *REDD1* significantly delayed AKT activation and suppressed cell proliferation. In *REDD1* knockdown cells, nitrates minimally influenced *REDD1* expression; however, cisplatin continued to inhibit OSCC growth. Blocking AKT activation with perifosine, a small-molecule inhibitor of AKT, significantly delayed pAKT expression. Blocking AKT activation with

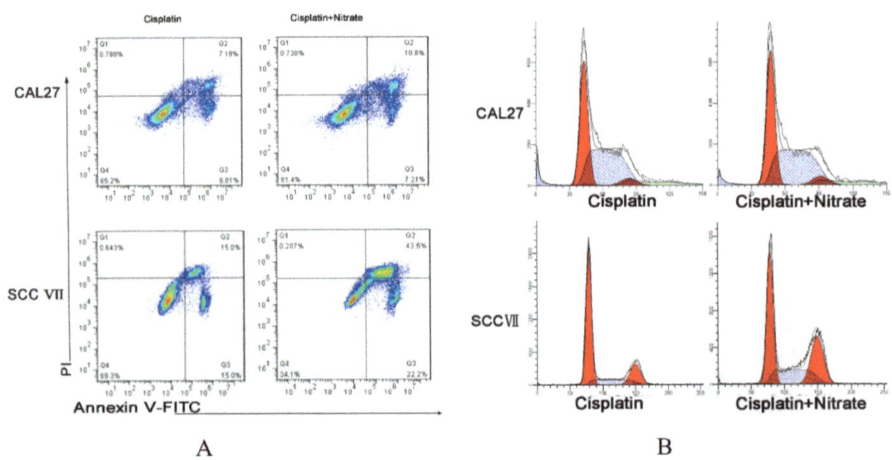

Fig. 8.2 The combined inhibitory effect of nitrate and cisplatin on OSCC proliferation. (**a**) Nitrate increased cisplatin-induced apoptosis. (**b**) The combination of cisplatin-nitrate increased cell cycle arrest

perifosine resulted in minimal impact of nitrates on the growth inhibition of OSCC by cisplatin. The findings indicate that nitrates augment the inhibitory effects of cisplatin on OSCC growth via the *REDD1*/AKT signaling pathway.

8.1.2.4 The Protective Role of Nitrate in Organs During Cisplatin Chemotherapy

To evaluate the protective effects of nitrate on organs during cisplatin chemotherapy, various organs, including the heart, liver, kidneys, stomach, intestine, and submandibular gland, were collected from mice following cisplatin treatment. The tissues underwent hematoxylin and eosin (HE) staining. The findings indicated that nitrate treatment significantly alleviated cisplatin-induced organ toxicity, especially in the heart, liver, and kidneys (Fig. 8.3). Furthermore, nitrate also exhibited protective effects against cisplatin-induced damage in the stomach, intestine, and submandibular gland.

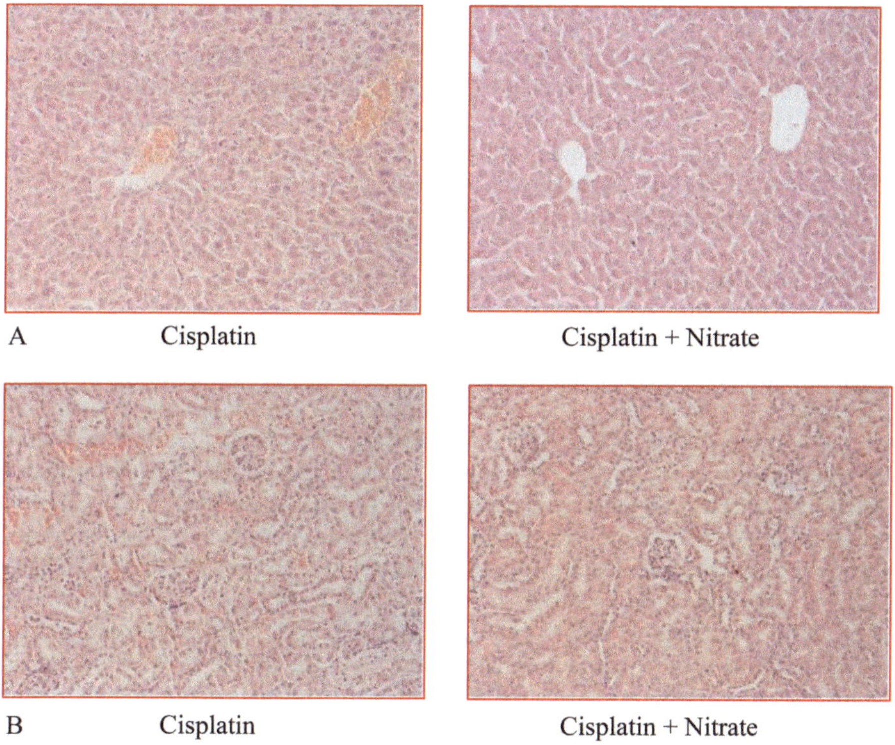

Fig. 8.3 Effect of nitrate application on hepatorenal organ toxicity induced by cisplatin. (**a**) Nitrate alleviated chemotherapy-induced hepatotoxicity, resulting in a decrease in steatosis of peripheral hepatic cells in certain lobules, disordered hepatic cord arrangement (80%), and hepatic congestion (10%). (**b**) Nitrate decreased congestion and edema in renal interstitial and tubular epithelial cells

8.1.3 Signaling Pathway of Sensitization Induced by Nitrate Chemotherapy

Both in vitro and in vivo studies indicate that nitrates contribute to chemosensitization under specific conditions, primarily through the inhibition of the *REDD1*/AKT signaling pathway.

8.1.3.1 REDD1 Modulates Chemotherapy Sensitivity Across Different Cancer Types

REDD1 plays a protective role in various cancer types by safeguarding cells against DNA damage induced by cisplatin. The overexpression of REDD1 has been demonstrated to diminish sensitivity to cisplatin chemotherapy in these cancers [14]. In ovarian cancer cells, the up-regulation of *REDD1* correlates with enhanced resistance to cisplatin [20]. Similarly, combining nitrate treatment with cisplatin enhances chemotherapy sensitivity in OSCC through the reduction of *REDD1* expression. This finding aligns with prior research that associated reduced cisplatin sensitivity with elevated REDD1 expression [24]. In OSCC cells subjected to cisplatin treatment, *REDD1* expression was found to be upregulated; however, nitrate treatment led to a significant reduction in this expression. *REDD1* is regulated by several transcription factors, such as P53, P63, activating transcription factor 4, SP1, and HIF-1 [25]. Nitrates produce NO, which modulates the function of P53 through conformational changes [26] and activates HIF-1 by stabilizing HIF-1α [27]. Although NO has been shown to modulate *REDD1* expression, the exact molecular mechanisms remain unclear [28].

8.1.3.2 The Role of REDD1's in Nitrate-Cisplatin Chemotherapy Sensitization

Hypoxia, a recognized characteristic of solid tumors, significantly contributes to resistance to chemotherapy and radiation therapy. In hypoxic conditions, cancer cells reduce oxidative metabolism and employ ATP protection mechanisms by restricting protein synthesis, which contributes to chemotherapy resistance, including resistance to cisplatin [29, 30]. Under hypoxic conditions, cisplatin-resistant gastric cancer cells demonstrate elevated levels of HIF-1α and pAKT. The targeted inhibition of AKT led to a decrease in HIF-1α expression, thereby increasing the sensitivity of cisplatin-resistant cells to cisplatin [31]. The activation of the PI3K/AKT/mTOR pathway induced by hypoxia has been associated with the promotion of cisplatin resistance in low-oxygen environments [32]. The overexpression of *REDD1* confers cellular protection under stress via the REDD1/AKT/mTOR signaling pathway. Reduced levels of *REDD1* sensitize cells to apoptosis, while increased *REDD1* expression confers protection against apoptotic stimuli [22]. Elevated *REDD1* levels can diminish cisplatin efficacy by downregulating the phosphorylation of 4E-BP1 and S6K, which are downstream targets of the mTOR pathway. The observed decrease in pAKT after *REDD1* suppression may also contribute to the increased sensitivity of hypoxic tumor cells to cisplatin through nitrate treatment.

Nitrates have been demonstrated to restore the sensitivity of cancer cells to cisplatin-induced apoptosis by inhibiting cell growth, intensifying cell cycle arrest, and diminishing *REDD1*'s protective effect. In the context of hypoxia, these effects may be associated with NO's capacity to inhibit the expression and transcriptional activity of HIF-1α or destabilize the HIF-1α protein, ultimately leading to reduced HIF-1α accumulation and increased chemosensitization. Nevertheless, the precise mechanisms governing these processes necessitate additional exploration in forthcoming experiments.

8.1.3.3 Nitrates and Chemotherapy Sensitization as an NO Donor

There is increasing evidence suggesting that dietary nitrates do not correlate with carcinogenic effects and may, in fact, decrease cancer incidence [33, 34]. Nitrates have diverse biological effects, including anti-tumor properties and the enhancement of chemotherapy sensitivity, mediated through various pathways and signaling cascades. Nitrates, as an NO donor, release significant amounts of NO, which contributes to anti-tumor effects and is essential for enhancing chemotherapy sensitization [35]. NO mitigates drug and immune resistance resulting from the dysregulation of the NF-κB/Snail/PTEN loop, either directly or indirectly, through the modulation of gene products.

Elevated levels of NO suppress NF-κB and its target gene RKIP (Raf kinase inhibitory protein), concurrently diminishing the inhibitor Snail. RKIP upregulation significantly inhibits the NF-κB and MAPK pathways, thereby enhancing the suppression of NF-κB, Snail, and YY1 expression, while concurrently promoting RKIP expression. This cascade results in decreased tumor cell proliferation, increased tumor cell apoptosis, and heightened sensitivity of cisplatin-resistant tumor cell lines to cisplatin chemotherapy [36]. PTEN, involved in the PI3K/AKT anti-apoptotic pathway, is subject to negative regulation by Snail and YY1. Activation of NF-κB, Snail, and YY1 in tumor cells leads to the inactivation of PTEN, subsequently activating the PI3K/AKT anti-apoptotic pathway and promoting tumor cell growth. Elevated levels of NO can lead to the up-regulation of PTEN, inhibition of the PI3K/AKT pathway, and suppression of downstream gene expression that facilitates tumor cell proliferation and anti-apoptotic processes. The effects augment the sensitization of tumor cells to chemotherapy (Fig. 8.4).

8.1.3.4 NO Interacts with Other Molecular Reaction Products to Enhance Chemotherapy Sensitization Effects

NO can react with other molecules, such as O_2^-, to produce nitrite (NO_2^-) and peroxynitrite ($ONOO^-$), which function as DNA oxidants and can cause single-strand breaks in DNA. Furthermore, nitrite can influence cyclic GMP (cGMP) production and inhibits the mitochondrial respiratory chain, which promotes the generation of reactive oxygen species (ROS). ROS and NO can generate reactive nitrogen species (RNS), which hinder NF-κB activation and directly interact with DNA or modify associated proteins. These alterations modulate downstream signaling pathways, ultimately triggering apoptosis and increasing sensitivity to chemotherapy [37] (Fig. 8.5).

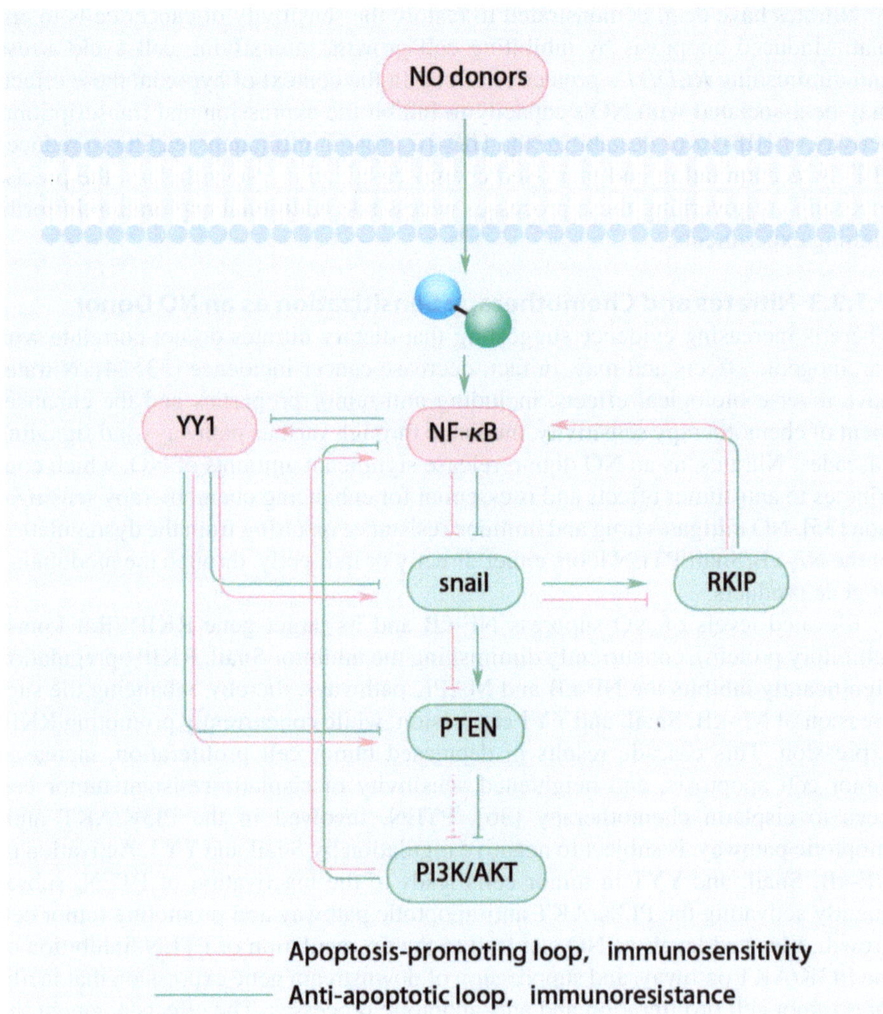

Fig. 8.4 NO donor reverses the anti-apoptotic NF-κB/Snail/YY1/RKIP/PTEN resistance loops into the pro-apoptotic loops of the chemotherapy sensitizing effect

8.1.3.5 The Protective Role of Nitrates in Chemotherapy-Induced Organ Toxicity

Chemotherapy agents are associated with hepatic impairment and can induce cardiotoxicity, nephrotoxicity, and neurotoxicity. Nitrates demonstrate protective effects against organ toxicity induced by chemotherapy. Supplementing exogenous nitrates can restore serum and liver nitrate levels, thereby aiding in the prevention of liver function decline. Furthermore, nitrates have the potential to mitigate damage to the heart, liver, and kidneys by decreasing blood pressure, enhancing microvascular density, and facilitating vasodilation [38].

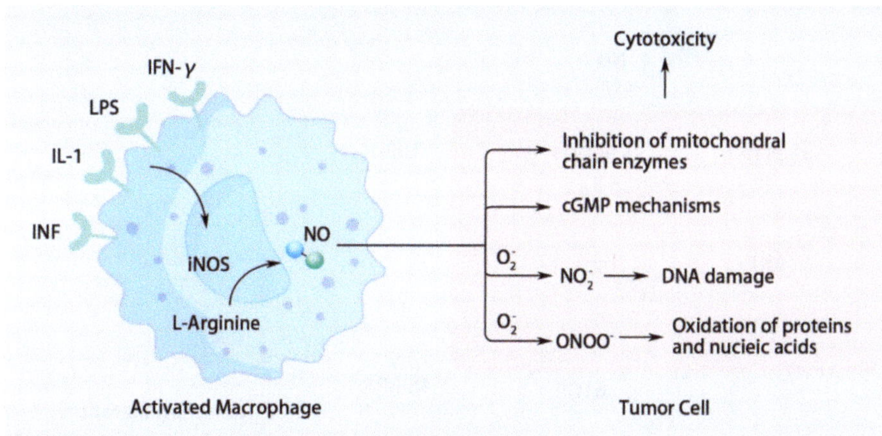

Fig. 8.5 NO synthesis and its impact on tumor cells

In summary, exogenous nitrate supplementation enhances chemotherapy sensitization and reduces the toxic side effects associated with chemotherapy drugs. However, the precise mechanisms underlying these effects necessitate additional investigation. *REDD1* and pAKT serve as critical regulators of cisplatin chemotherapy sensitivity in OSCC cells. Increased *REDD1* expression may stimulate AKT, while nitrates may improve cisplatin sensitivity in OSCC cells by decreasing *REDD1* expression and reducing AKT activation. Based on prior research, the hypothesis is that the mechanism may involve the inhibition of HIF-1α expression and the activation of the TSC/mTOR/4E-BP1/S6K pathway, resulting in decreased tumor cell proliferation and enhanced apoptosis. Therefore, reducing *REDD1* expression through nitrate supplementation may offer a novel approach to improve sensitivity to cisplatin chemotherapy (Fig. 8.6).

8.2 The Role and Mechanism of Sialin Protein in the Development and Progression of Thyroid Cancer

Haibo Wang

8.2.1 Expression of Sialin in Thyroid Cancer Tissue

Sialin protein is present in multiple human tissues, including the central nervous system, gastrointestinal tract, and kidneys, with peak expression observed in endocrine organs like the thyroid. In thyroid cancer, sialin expression levels are two to three times greater than those in adjacent noncancerous tissues. Moreover, in patients with thyroid cancer possessing the BRAFv600e mutation, mRNA levels of

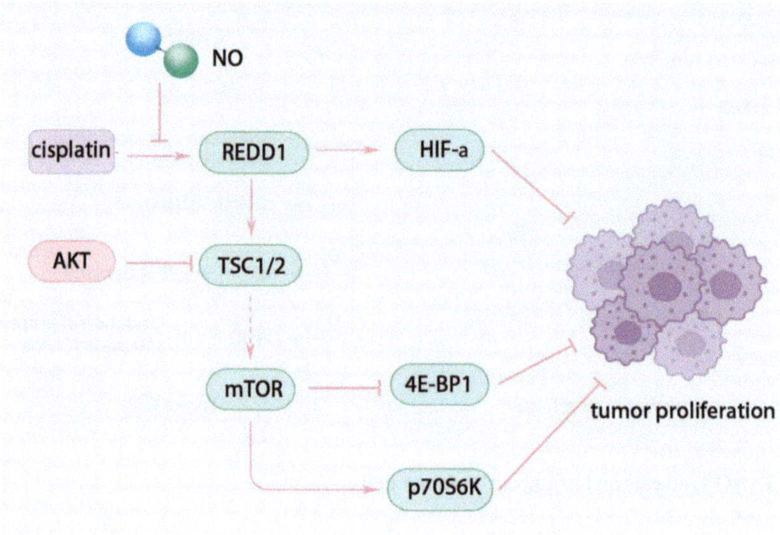

Fig. 8.6 The mechanism through which nitrates may enhance chemotherapy sensitization to cisplatin via NO

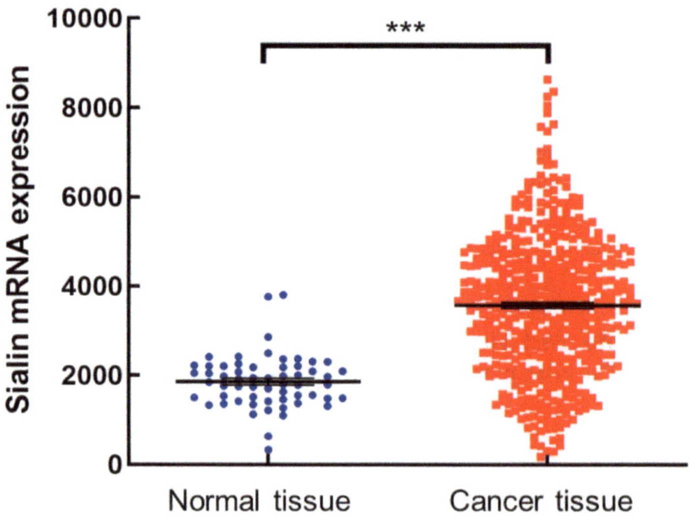

Fig. 8.7 The mRNA expression levels of sialin in thyroid tumors and adjacent tissues

sialin are significantly higher compared to those with the wild-type BRAF (Fig. 8.7). The upregulation of sialin at the mRNA level is highly associated with the BRAFv600e mutation.

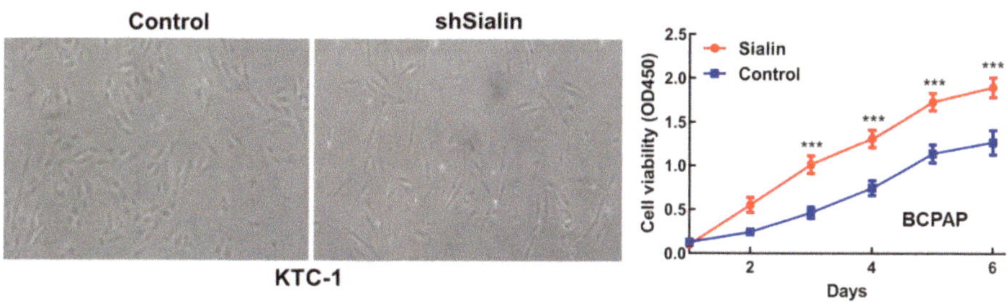

Fig. 8.8 The proliferation capabilities of thyroid cancer cell lines with stable knockdown or overexpression of sialin

8.2.2 The Biological Role of Sialin in Thyroid Cancer Cells

Knocking down sialin in the KTC-1 thyroid cancer cell line led to a huge decrease in cell proliferation and colony-forming capacity. In contrast, the overexpression of sialin in the BCPAP thyroid cancer cell line increased both proliferation and colony-forming abilities (Fig. 8.8). In vivo experiments utilizing these thyroid cancer cell models demonstrated that the knockdown of sialin resulted in inhibited tumor growth, while the overexpression of sialin facilitated tumor growth.

8.2.3 Molecular Mechanism of Sialin in Promoting Thyroid Cancer Proliferation

Research into the associated mechanisms indicates that the knockdown of sialin in thyroid cancer cells significantly suppressed the expression of the cell cycle protein CCND1, implying that the lack of sialin results in cell cycle arrest. Subsequent experiments indicated that the downregulation of sialin hindered mTOR pathway activity (Fig. 8.9). The mTOR pathway is modulated by a range of cellular signals, including mitogenic growth factors, hormones like insulin, nutrients such as amino acids and glucose, cellular energy status, and stress conditions. The absence of sialin inhibited the activity of the mTOR pathway in thyroid cancer cells, even under conditions of starvation or stimulation by amino acids alone. This indicates that in the advancement of thyroid cancer, elevated levels of sialin may counteract the negative effects of nutrient deficiency, thereby facilitating continuous proliferation and progression of the disease.

8.3 Intestinal Tumor Prevention and Treatment

Lei Hu

Colorectal cancer is a prevalent malignant tumor of the intestinal tract, with increasing morbidity and mortality rates observed annually, making it a significant contributor to human mortality [39, 40]. The pathogenesis of colorectal cancer is

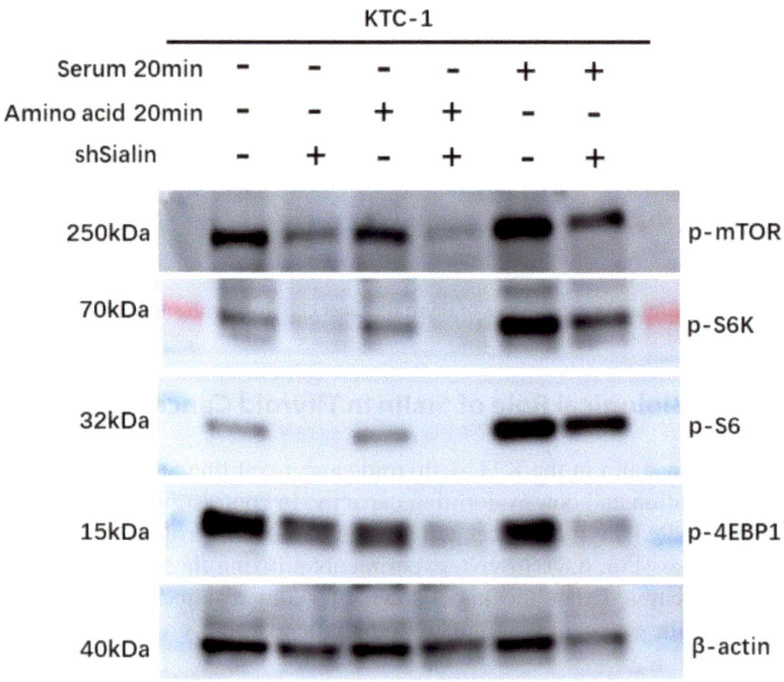

Fig. 8.9 The effect of sialin knockdown on mTOR pathway activity in KTC-1 cells across various nutritional conditions, as assessed by Western blot analysis

complex, arising from an interplay of genetic and environmental factors [41–43]. The primary risk factors include disorders of intestinal microecology and the intestinal mucosal immune system, excessive consumption of red and processed meats, smoking, alcohol consumption, and obesity. The primary treatment modalities for colorectal cancer encompass endoscopic treatment, surgical resection, radiotherapy, chemotherapy, targeted therapy, immunotherapy, etc. However, each treatment modality presents limitations, including the significant trauma associated with surgical interventions, the potential for tumor resistance to chemotherapeutic agents, and the occurrence of lymph node or distal metastasis at diagnosis due to the absence of prominent early stage symptoms [44–48]. Consequently, identifying effective pharmacological agents to decrease the occurrence of intestinal tumors is essential.

Dietary nitrate, a natural component of food, is absorbed into the bloodstream via the intestinal mucosa and subsequently reabsorbed by the salivary glands, where it is secreted into saliva and delivered to the gastrointestinal tract, establishing a salivary gland-intestinal cycle [49]. Prior epidemiological research has identified nitrate as a potential carcinogen, and numerous studies have investigated the relationship between nitrate, nitrite, and tumorigenesis. To date, epidemiological evidence of the association between nitrate in drinking water and the risk of colorectal cancer remains limited [50–53]. In a mouse model of colon cancer induced by azomethane and dextran sodium sulfate (AOM/DSS), the research group observed a significant reduction in body weight, decreased fecal continuity, and positive fecal occult blood test results in both control and nitrate groups of mice following the prophylactic addition of nitrate to drinking water. Mice in the nitrate group exhibited a significant decrease in both the number and size of colon tumors when compared to the control group. This indicates that nitrate can significantly reduce the occurrence and development of colon tumors.

Histological analysis of mouse colon tumor tissues indicated that cellular morphology heterogeneity within colon tumors was diminished in the nitrate group compared to the control group, and the quantity of positive cells exhibiting proliferative capacity in tumor tissues was significantly lower. Detection of microbes in the colon via 16sRNA revealed that nitrate enhanced the abundance of dominant bacteria, specifically *Bacteroidetes* and *Proteobacteria*, in the intestines of healthy individuals, while reducing the abundance of *Firmicutes* and *Fusobacteria* in patients with colorectal cancer. Furthermore, neutrophils serve as a reliable prognostic marker for colorectal cancer treatment and exhibit a significant correlation with patient survival. Neutrophils frequently aggregate around and within tumors, contributing to tumor growth, and these neutrophils are referred to as tumor-associated neutrophils (TAN). In contrast to conventional neutrophils, tumor-associated neutrophils exhibit more intensive immunological activity and possess an extended maturation and lifespan, positioning them as essential components of the tumor immune microenvironment. The research group discovered that nitrates influence the immune microenvironment in mouse colon tumors through the modulation of myeloid lymphocytes, including neutrophils.

Nitrates also preserve the homeostasis of intestinal epithelial cells during tumor induction by inhibiting the function and proliferation of intestinal stem cells and intestinal crypt stem cells. A mouse intestinal organoid model demonstrated that nitrates support organoid growth and inhibit tumor proliferation. Nitrates simultaneously up-regulate apoptosis, ferroptosis, oxidative stress levels, and the expression of immunotherapy checkpoints such as PD1, indicating its potential to enhance the therapeutic effects of tumor therapy (Fig. 8.10).

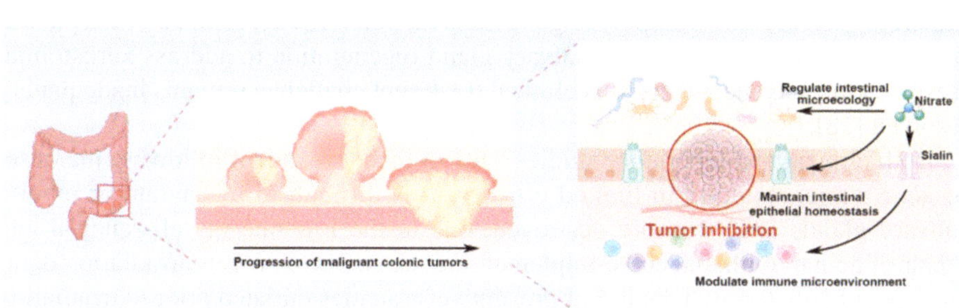

Fig. 8.10 Mechanism of nitrates in the intestinal tumor prevention and treatment

8.4 Prevention of Irradiation-Induced Damage

Xiaoyu Feng

(i) Overview of irradiation-induced xerostomia.

Head and neck cancer (HNC) exhibits a significant global incidence [54], frequently requiring radiotherapy. Salivary glands in the anatomical region of the head and neck are susceptible to irradiation damage, which presents as irradiation-induced xerostomia [55]. Xerostomia induced by irradiation is a prevalent complication among head and neck cancer patients undergoing radiotherapy, leading to increased caries and mucositis. Consequently, the quality of life for these patients is markedly impaired [56]. Patients with HNC began to show signs of xerostomia 1 week post-irradiation therapy, with the symptom progressively worsening over time. Follow-up indicated no alleviation of irradiation-induced xerostomia at two or more years after irradiation therapy [56].

The alveolar epithelium of salivary glands exhibits high radiosensitivity, whereas the ductal epithelium demonstrates relative non-radiosensitivity. Two types of alveolar epithelial cells exist: serous and mucinous. Serous alveolar epithelial cells exhibit high radiosensitivity, whereas mucinous alveolar epithelial cells demonstrate relative non-radiosensitivity [57]. Serous epithelial cells are highly differentiated with an extended average lifespan and reduced mitotic activity, suggesting a theoretical non-radiosensitivity. However, clinical observations indicate that purely serous glands, such as the parotid gland, and mixed glands like the submandibular gland, which are acutely responsive to radiation, exhibit dry mouth symptoms shortly after radiotherapy [57]. Histopathological examination indicates significant glandular atrophy, extensive loss of vesicular structures, and pronounced fibrous tissue infiltration [56].

Currently, artificial saliva and pilocarpine are frequently employed to mitigate the symptoms of irradiation-induced xerostomia; however, their efficacy is limited, and no effective treatment for irradiation-induced xerostomia exists at this time. Laboratory studies have employed gene transduction (AQP1), stem cell transplantation, immunosuppressants, and salivary gland regeneration to address xerostomia. Despite some advancements, this clinical treatment challenge remains inadequately resolved [58].

Both clinical and laboratory studies have initiated treatment following the onset of xerostomia, resulting in limited efficacy. It is difficult to regenerate or rebuild salivary glands after they are destroyed, due to the low rate of self-renewal and repair of acinar epithelial cells. Implementing measures to prevent irradiation damage, specifically through the use of preventive strategies initiated prior to irradiation therapy, can significantly preserve the histological structure and secretory function of the salivary glands. This represents a novel foundation for future research focused on the prevention and treatment of xerostomia.

(ii) Nitrates for the prevention and treatment of irradiation induced xerostomia

Upon absorption into the bloodstream via the gastrointestinal mucosa, nitrates are actively transported and secreted into saliva by sialin, resulting in a saliva concentration of nitrate that is tenfold higher than that in the blood. Sialin functions as a nitrate transport channel in the salivary glands, exhibiting the highest expression levels in this tissue [59, 60]. Nonetheless, the physiological significance of sialin-mediated active transport of nitrates in salivary glands remains unclear. The nitrate-sialin feedback loop is hypothesized to play a role in maintaining the normal physiological function of salivary glands, suggesting that nitrates may be beneficial for the prevention and treatment of irradiation-induced xerostomia.

1. Dietary nitrate supplementation may mitigate irradiation-induced damage to the salivary glands. The laboratory of Songlin Wang previously developed an animal model to study radiation injury in the parotid gland of mini-pigs, simulating early stage clinical head and neck cancer split radiotherapy. This model involved irradiating the unilateral parotid gland with a single dose of 7.5 Gy for seven consecutive days, resulting in a total biological effect of 58 Gy, comparable to the clinical therapeutic dose [61]. Following this model, exogenous oral supplementation with varying doses of inorganic nitrates (2, 1, 0.5, 0.25 mmol/kg-day) commenced 1 week prior to segmented radiotherapy. One week later, the unilateral parotid glands of miniature pigs underwent segmented irradiation, with continuous nitrate administration for 4 months until the observation endpoints. The study aimed to determine whether nitrates could mitigate radiation damage to salivary glands and whether their protective effect was dose-dependent. Throughout the observation cycle, the research group determined that the highest nitrate dose (2 mmol/kg-day) effectively maintained parotid gland salivary secretion comparable to the normal control. This ability was preserved at approximately 85% until the conclusion of the 4-month observation period following radiotherapy. Additionally, salivary secretion in the other nitrate dosage groups (1, 0.5, 0.25 mmol/kg-day) was 65%, 50%, and 30% of pre-radiotherapy levels, respectively. In contrast, the radiotherapy control group exhibited a salivary flow rate of approximately 15% of pre-radiotherapy levels. The results indicate that nitrate prophylactic administration effectively protects salivary glands from radiation damage and prevents the onset of radioactive dry mouth. The performance of parotid surface blood flow correlated with the salivary flow rate, demonstrating significant dose dependence, with improved surface blood flow observed at high doses of nitrates. Four months post-radiotherapy, animals were euthanized, and glandular tissues were weighed. A significant dose-dependent relationship was observed between gland weight and the administered nitrate dose. Specifically, higher doses correlated with gland weights approaching those of normal parotid glands. In contrast, 2 months post-radiotherapy, following the establishment of radiation damage during the mid to late stages of treatment, exogenous nitrate supplementation did not demonstrate a protective effect. This indicates that only prophylactic administration of nitrates can effectively safeguard the secretory function of the parotid gland [61].

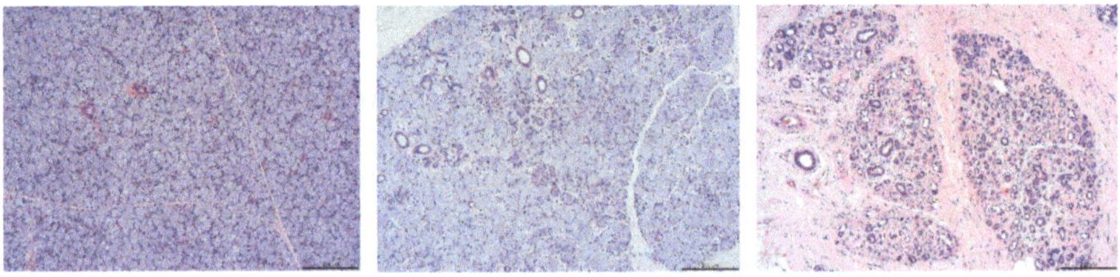

Fig. 8.11 Dietary nitrate supplementation effectively protects salivary gland from irradiation damage

Histological examination indicated that in the parotid tissue of the radiotherapy control group, a large number of alveolar cells and lobules disappeared, with ducts exhibiting irregular curvature, dilation, or atresia. Additionally, the fibrous tissues displayed extensive infiltration, characteristic of parotid radiation injury. In contrast, the morphology of parotid tissue treated with 2 mmol/kg-day nitrate resembled that of normal parotid tissue, exhibiting intact glandular follicles and lobules, with vacuolar changes noted in only a minimal number of glandular follicular cells under high magnification microscopy. As nitrate administration dosage decreased, there was a corresponding reduction in the number of follicular cells and an increase in fibrous tissue infiltration, demonstrating a clear dose-dependent relationship. This was consistent with observed changes in salivary flow rate, surface blood flow, and glandular weight (Fig. 8.11). Additionally, Masson collagen fiber staining results indicated that higher nitrate doses were associated with reduced fibrous tissue infiltration, with histological findings resembling those of normal tissue. Immunohistochemical staining results indicated that nitrates promoted the proliferation of adenohypophysis (ki67), microvessel formation in the tissue (CD31), and the expression of sialin [62].

The study results regarding the use of nitrates to prevent radiological dry mouth were published as a cover article in eLife in 2021 [61]. Professors Harry Quon and Fred Bunz from the Department of Radiation Oncology and Radiomolecular Sciences at Johns Hopkins University School of Medicine, Baltimore, USA, provided commentary on this paper in eLife. They evaluated the study's findings [62], noting that the administration of inorganic nitrates can prevent xerostomia. They highlighted the simplicity of inorganic nitrates in terms of composition and administration, the potential to address barriers in translating results from animal studies to human applications, and the characterization of this approach as a low-tech solution to a high-tech issue.

In conclusion, inorganic nitrates are widely available in water and food, and the World Health Organization has established that nitrates are not carcinogenic [63]. Inorganic nitrates are administered orally, exhibiting high absorption efficiency, absence of tolerance, lack of immunogenicity, and no toxic side effects. It is safe and easy to administer, presenting significant potential as a novel prevention and treatment modality for salivary gland radiation injury in future clinical applications.

Compliance with Ethical Standards All applicable international, national, and/or institutional guidelines for the care and use of animals were followed.

References

1. Thavarool SB, Muttath G, Nayanar S, et al. Improved survival among oral cancer patients: findings from a retrospective study at a tertiary care cancer Centre in rural Kerala, India. World J Surg Oncol. 2019;17(1):15.
2. Ghani WMN, Ramanathan A, Prime SS, et al. Survival of oral cancer patients in different ethnicities. Cancer Investig. 2019;37(7):275–87.
3. Zhong LP, Zhang CP, Ren GX, et al. Randomized phase III trial of induction chemotherapy with docetaxel, cisplatin, and fluorouracil followed by surgery versus up-front surgery in locally advanced resectable oral squamous cell carcinoma. J Clin Oncol. 2013;31(6):744–51.
4. Vermorken JB, Mesia R, Rivera F, et al. Platinum-based chemotherapy plus cetuximab in head and neck cancer. N Engl J Med. 2008;359(11):1116–27.
5. Huang ZX, Zhang Y, Li HG, et al. Vitamin D promotes the cisplatin sensitivity of oral squamous cell carcinoma by inhibiting LCN2-modulated NF-κB pathway activation through RPS3. Cell Death Dis. 2019;10(12):936.
6. Tanzawa H, Uzawa K, Kasamatsu A, et al. Targeting gene therapies enhance sensitivity to chemo- and radiotherapy of human oral squamous cell carcinoma. Oral Sci Int. 2015;12(2):43–52.
7. Lundberg JO, Govoni M. Inorganic nitrate is a possible source for systemic generation of nitric oxide. Free Radic Biol Med. 2004;37(3):395–400.
8. Ning SC, Bednarski M, Oronsky B, et al. Novel nitric oxide generating compound glycidyl nitrate enhances the therapeutic efficacy of chemotherapy and radiotherapy. Biochem Biophys Res Commun. 2014;447(3):537–42.
9. Frederiksen LJ, Sullivan R, Maxwell LR, et al. Chemosensitization of cancer *in vitro* and *in vivo* by nitric oxide signaling. Clin Cancer Res. 2007;13(7):2199–206.
10. Rishi L, Dhiman R, Raje M, et al. Nitric oxide induces apoptosis in cutaneous T cell lymphoma (HuT-78) by downregulating constitutive NF-kappaB. Biochim Biophys Acta. 2007;1770(8):1230–9.
11. Azizzadeh B, Yip HT, Blackwell KE, et al. Nitric oxide improves cisplatin cytotoxicity in head and neck squamous cell carcinoma. Laryngoscope. 2001;111(11 Pt 1):1896–900.
12. Perrotta C, Bizzozero L, Falcone S, et al. Nitric oxide boosts chemoimmunotherapy via inhibition of acid sphingomyelinase in a mouse model of melanoma. Cancer Res. 2007;67(16):7559–64.
13. Stanimirovic J, Obradovic M, Jovanovic A, et al. A high fat diet induces sex-specific differences in hepatic lipid metabolism and nitrite/nitrate in rats. Nitric Oxide. 2016;54:51–9.
14. Jin HO, Hong SE, Kim JH, et al. Sustained overexpression of Redd1 leads to Akt activation involved in cell survival. Cancer Lett. 2013;336(2):319–24.
15. Gong JL, Zhou SX, Yang SH. Vanillic acid suppresses HIF-1alpha expression via inhibition of mTOR/p70S6K/4E-BP1 and Raf/MEK/ERK pathways in human colon cancer HCT116 cells. Int J Mol Sci. 2019;20(3):465.
16. Shoshani T, Faerman A, Mett I, et al. Identification of a novel hypoxia-inducible factor 1-responsive gene, RTP801, involved in apoptosis. Mol Cell Biol. 2002;22(7):2283–93.
17. McGhee NK, Jefferson LS, Kimball SR. Elevated corticosterone associated with food deprivation upregulates expression in rat skeletal muscle of the mTORC1 repressor, REDD1. J Nutr. 2009;139(5):828–34.
18. Li XH, Ha CT, Fu D, et al. REDD1 protects osteoblast cells from gamma radiation-induced premature senescence. PLoS One. 2012;7(5):e36604.

19. Lee SY, Kang HG, Yoo SS, et al. Polymorphisms in DNA repair and apoptosis-related genes and clinical outcomes of patients with non-small cell lung cancer treated with first-line paclitaxel-cisplatin chemotherapy. Lung Cancer. 2013;82(2):330–9.
20. Chang B, Meng J, Zhu HM, et al. Overexpression of the recently identified oncogene REDD1 correlates with tumor progression and is an independent unfavorable prognostic factor for ovarian carcinoma. Diagn Pathol. 2018;13(1):87.
21. Dennis MD, NK MG, Jefferson LS, et al. Regulated in DNA damage and development 1 (REDD1) promotes cell survival during serum deprivation by sustaining repression of signaling through the mechanistic target of rapamycin in complex 1 (mTORC1). Cell Signal. 2013;25(12):2709–16.
22. Schwarzer R, Tondera D, Arnold W, et al. REDD1 integrates hypoxia-mediated survival signaling downstream of phosphatidylinositol 3-kinase. Oncogene. 2005;24(7):1138–49.
23. Guo GF, Wang YX, Zhang YJ, et al. Predictive and prognostic implications of 4E-BP1, Beclin-1, and LC3 for cetuximab treatment combined with chemotherapy in advanced colorectal cancer with wild-type KRAS: analysis from real-world data. World J Gastroenterol. 2019;25(15):1840–53.
24. Zeng QH, Liu JY, Cao PG, et al. Inhibition of REDD1 sensitizes bladder urothelial carcinoma to paclitaxel by inhibiting autophagy. Clin Cancer Res. 2018;24(2):445–59.
25. Whitney ML, Jefferson LS, Kimball SR. ATF4 is necessary and sufficient for ER stress-induced upregulation of REDD1 expression. Biochem Biophs Res Commun. 2009;379(2):451–5.
26. Chang SM, Hu L, Xu YP, et al. Inorganic nitrate alleviates total body irradiation-induced systemic damage by decreasing reactive oxygen species levels. Int J Radiat Oncol Biol Phys. 2019;103(4):945–57.
27. Lee JY, Hirota SA, Glover LE, et al. Effects of nitric oxide and reactive oxygen species on HIF-1α stabilization following clostridium difficile toxin exposure of the Caco-2 epithelial cell line. Cell Physiol Biochem. 2013;32(2):417–30.
28. Turpaev K, Glatigny A, Bignon J, et al. Variation in gene expression profiles of human monocytic U937 cells exposed to various fluxes of nitric oxide. Free Radic Biol Med. 2010;48(2):298–305.
29. Ma T, Patel H, Babapoor-Farrokhran S, et al. KSHV induces aerobic glycolysis and angiogenesis through HIF-1-dependent upregulation of pyruvate kinase 2 in Kaposi's sarcoma. Angiogenesis. 2015;18(4):477–88.
30. Ostergaard L, Tietze A, Nielsen T, et al. The relationship between tumor blood flow, angiogenesis, tumor hypoxia, and aerobic glycolysis. Cancer Res. 2013;73(18):5618–24.
31. Sun XP, Dong XS, Lin L, et al. Up-regulation of surviving by AKT and hypoxia-inducible factor 1α contributes to cisplatin resistance in gastric cancer. FEBS J. 2014;281(1):115–28.
32. Gong TX, Cui LQ, Wang HL, et al. Knockdown of KLF5 suppresses hypoxia-induced resistance to cisplatin in NSCLC cells by regulating HIF-1α-dependent glycolysis through inactivation of the PI3K/Akt/mTOR pathway. J Transl Med. 2018;16(1):164.
33. Bryan NS, Alexander DD, Coughlin JR, et al. Ingested nitrate and nitrite and stomach cancer risk: an updated review. Food Chem Toxicol. 2012;50(10):3646–465.
34. Jakszyn P, Gonzalez CA. Nitrosamine and related food intake and gastric and oesophageal cancer risk: a systematic review of the epidemiological evidence. World J Gastroenterol. 2006;12(27):4296–303.
35. Bian K, Murad F. sGC-cGMP signaling: target for anticancer therapy. Adv Exp Med Biol. 2014;814:5–13.
36. Bonavida B, Garban H. Nitric oxide-mediated sensitization of resistant tumor cells to apoptosis by chemo-immunotherapeutics. Redox Biol. 2015;6:486–94.
37. Li WW, Han WX, Ma YM, et al. P53-dependent miRNAs mediate nitric oxide-induced apoptosis in colonic carcinogenesis. Free Radic Biol Med. 2015;85:105–13.
38. Carlström M, Persson AE, Larsson E, et al. Dietary nitrate attenuates oxidative stress, prevents cardiac and renal injuries, and reduces blood pressure in salt-induced hypertension. Cardiovasc Res. 2011;89(3):574–85.
39. Dekker E, Tanis PJ, Vleugels JLA, et al. Colorectal cancer. Lancet. 2019;394(10207):1467–80.

40. Jung G, Hernández-Illán E, Moreira L, et al. Epigenetics of colorectal cancer: biomarker and therapeutic potential. Nat Rev Gastroenterol Hepatol. 2020;17(2):111–30.
41. Coker OO, Nakatsu G, Dai RZ, et al. Enteric fungal microbiota dysbiosis and ecological alterations in colorectal cancer. Gut. 2019;68(4):654–62.
42. Henrikson NB, Webber EM, Goddard KA, et al. Family history and the natural history of colorectal cancer: systematic review. Genet Med. 2015;17(9):702–12.
43. Schoen RE, Razzak A, Yu KJ, et al. Incidence and mortality of colorectal cancer in individuals with a family history of colorectal cancer. Gastroenterology. 2015;149(6):1438–1445.e1.
44. Jayanna M, Burgess NG, Singh R, et al. Cost analysis of endoscopic mucosal resection vs surgery for large laterally spreading colorectal lesions. Clin Gastroenterol Hepatol. 2016;14(2):271–18.
45. Bondeven P, Hagemann-Madsen RH, Laurberg S, et al. Extent and completeness of mesorectal excision evaluated by postoperative magnetic resonance imaging. Br J Surg. 2013;100(10):1357–67.
46. Du DL SZR, Wang D, et al. Optimal interval to surgery after neoadjuvant chemoradiotherapy in rectal cancer: a systematic review and meta-analysis. Clin Colorectal Cancer. 2018;17(1):13–24.
47. André T, Boni C, Navarro M, et al. Improved overall survival with oxaliplatin, fluorouracil, and leucovorin as adjuvant treatment in stage II or III colon cancer in the MOSAIC trial. J Clin Oncol. 2009;27(19):3109–16.
48. Hurwitz H, Fehrenbacher L, Novotny W, et al. Bevacizumab plus irinotecan, fluorouracil, and leucovorin for metastatic colorectal cancer. N Engl J Med. 2004;350(23):2335–42.
49. Pannala AS, Mani AR, Spencer JP, et al. The effect of dietary nitrate on salivary, plasma, and urinary nitrate metabolism in humans. Free Radic Biol Med. 2003;34(5):576–84.
50. Dellavalle CT, Xiao Q, Yang G, et al. Dietary nitrate and nitrite intake and risk of colorectal cancer in the Shanghai Women's Health Study. Int J Cancer. 2014;134(12):2917–26.
51. Chambers T, Douwes J, Mannetje A, et al. Nitrate in drinking water and cancer risk: the biological mechanism, epidemiological evidence and future research. Aust N Z J Public Health. 2022;46(2):105–8.
52. Kobayashi J. Effect of diet and gut environment on the gastrointestinal formation of N-nitroso compounds: a review. Nitric Oxide. 2018;73:66–73.
53. LRRA C, Guimarães DD, AFL F, et al. Effects of chronic dietary nitrate supplementation on longevity, vascular function and cancer incidence in rats. Redox Biol. 2021;48:102209.
54. Sung H, Ferlay J, Siegel RL, et al. Global cancer statistics 2020: GLOBOCAN estimates of incidence and mortality worldwide for 36 cancers in 185 countries. CA Cancer J Clin. 2021;71(3):209–49.
55. Jensen SB, Vissink A, Limesand KH, et al. Salivary gland hypofunction and xerostomia in head and neck radiation patients. J Natl Inst Monogr. 2019;53:lgz016.
56. Jasmer KJ, Gilman KE, Muñoz Forti K, et al. Radiation-induced salivary gland dysfunction: mechanisms, therapeutics and future directions. J Clin Med. 2020;9(12):4095.
57. Pedersen AML, Sørensen CE, Proctor GB, et al. Salivary secretion in health and disease. J Oral Rehabil. 2018;45(9):730–46.
58. Mercadante V, Al Hamad AA, Lodi G, et al. Interventions for the management of radiotherapy-induced xerostomia and hyposalivation: a systematic review and meta-analysis. Oral Oncol. 2017;66:64–74.
59. Xia DS, Deng DJ, Wang SL. Destruction of parotid glands affects nitrate and nitrite metabolism. J Dent Res. 2003;82(2):101–5.
60. Qin L, Liu X, Sun Q, et al. Sialin (SLC17A5) functions as a nitrate transporter in the plasma membrane. Proc Natl Acad Sci USA. 2012;109(33):13434–9.
61. Feng XY, Wu ZF, Xu JJ, et al. Dietary nitrate supplementation prevents radiotherapy-induced xerostomia. eLife. 2021;10:e70710.
62. Quon H, Bunz F. Preventing collateral damage. eLife. 2021;10:e74319.
63. Omar SA, Artime E, Webb AJ. A comparison of organic and inorganic nitrates/nitrites. Nitric Oxide. 2012;26(4):229–40.

Protection Against Diseases of the Nervous System

9

Yuhua Lei Pan

9.1 Prevalence

Nitrates are crucial for the maintenance of homeostasis. Previous studies have indicated potential adverse effects on the organism; however, Songlin Wang's group discovered that exogenous nitrate supplementation can prevent and control oral and systemic diseases [1]. Additionally, it plays a role in regulating various functions, including vascular tone, antioxidant activity, inhibition of inflammatory factor release, glucose metabolism regulation, and prevention of ischemia-reperfusion injury [2]. Neurological disease protection is an important part of the homeostatic regulation by nitrates.

9.2 The Role of Nitrate/NO in the Nervous System

9.2.1 NO Is a Double-Edged Sword in the Nervous System

During the mid-twentieth century, it was widely recognized that acetylcholine could induce vasodilation [3], and vasodilation is a process that necessitates the concurrent presence of vascular endothelial cells. In 1987, scientists discovered that vascular endothelial cells triggered vasodilation through the release of NO in response to acetylcholine stimulation [4].

NO is extensively present in the tissues of living organisms, particularly within neural tissues. It serves as a crucial messenger in the central nervous system, diffusing to adjacent peripheral neurons such as presynaptic nerve endings and astrocytes. NO binds to guanylyl cyclase (GC), thereby modulating synaptic signaling through

Y. L. Pan (✉)
Neurology Department, Shanghai East Hospital, School of Medicine, Tongji University, Shanghai, China

S. Wang (ed.), *Inorganic Nitrate and Human Body Homeostasis*,
https://doi.org/10.1007/978-981-95-2824-0_9

cGMP-mediated pathways, which facilitates blood vessel dilation relevant to cerebrovascular disease treatment [5, 6].

NO is regarded as a "double-edged sword," possessing both neuroprotective and neurotoxic properties. Previous research has identified NO as a potentially harmful molecule that can increase oxidative stress, resulting in neuronal loss and contributing to central nervous system paralysis and spasms. Conversely, it has been suggested that NO, at physiological concentrations, can facilitate long-term potentiation (LTP) associated with learning and memory [7], and may also provide anti-apoptotic or pro-survival effects against various neurotoxic responses through multiple mechanisms. In the peripheral nervous system, NO functions as a non-cholinergic, non-adrenergic neurotransmitter or mediator involved in nociceptive afferents and sensory transmission.

9.2.2 Nitrates Serve as a Significant Alternative Pathway for NO Synthase Under Ischemic Conditions

Nitrites and nitrates are typically regarded as metabolites of the noble gas NO. Recent research indicates that nitrates can be converted to nitrites in vivo following exogenous ingestion. Inorganic nitrites serve as a regulator of physiological functions and tissue responses to ischemia, while the more stable nitrate anion (NO_3^-), often considered biologically inert, is ultimately recycled to produce the classical signaling molecule, NO, representing an important alternative pathway to the traditional nitric oxide synthase (NOS) [8] (Fig. 9.1).

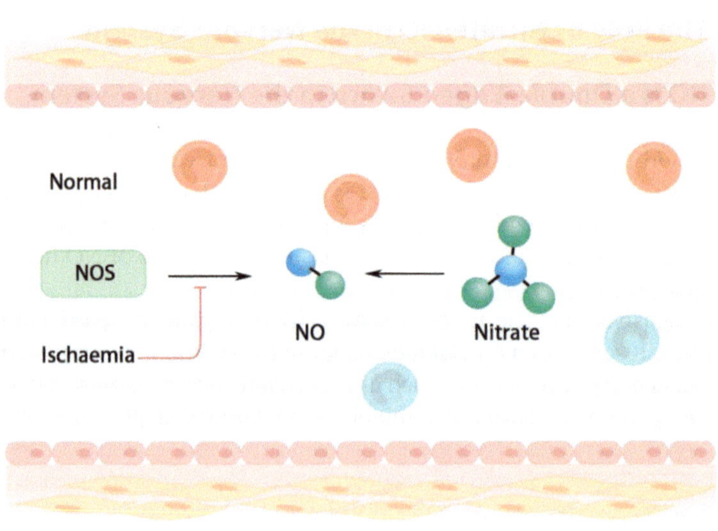

Fig. 9.1 Nitrates serve as a significant alternative pathway for NOS during ischemia and hypoxia, thereby ensuring NO production

The pathway of NO formation from nitrates is significantly enhanced in myocardial, neural, and other tissues, particularly under ischemic and hypoxic conditions. This enhancement protects NO production despite impaired oxygen-dependent NOS activity. The formation process inhibits mitochondrial respiration and the generation of mitochondria-derived reactive oxygen species, thereby contributing to the regulation of physiological hypoxic blood flow, etc. [9]. Nitrates can also enter cells through the sialin transporter, facilitating the nitrate/nitrite-NO pathway and maintaining the homeostatic NO cycle. Brain tissue exhibits a high expression level of the sialin transporter, second only to salivary glands, indicating a potential link between nitrate's role in the brain and sialin [10].

9.2.3 Types of NO Synthases

The human body contains three primary types of NO synthase: (1) endothelial nitric oxide synthase (eNOS), located in the endothelial cells of blood vessels; (2) neurogenic nitric oxide synthase (nNOS), found in neuronal cells; and (3) inducible nitric oxide synthase (iNOS), present in immune cells, including lymphocytes and T cells.

Multiple structural isoforms of NOS are prevalent in neurons, astrocytes, brain endothelial cells, and other cell types, facilitating the conversion of L-arginine to NO. Endothelial dysfunction and associated cellular damage in cerebrovascular disease impact the activity of eNOS and neuronal nNOS [11].

9.2.4 Nitrates and Cerebral Blood Flow

NO regulates physiological ischemic blood flow; however, the regulation of eNOS in the cerebral vasculature is more complex. Research utilizing eNOS inhibitors indicates an increase in total peripheral vascular resistance, yet this endothelial source of NO exhibits variable effects across different vessels. Limited information exists regarding the direct impact of nitrates or nitrites on the cerebral vasculature and their underlying mechanisms. A recent study demonstrated no effect of inorganic nitrate or nitrite supplementation on cerebral blood flow through the analysis of randomized controlled trials [12]. Further research is needed to unveil the potential influence of inorganic nitrate and nitrite supplementation on cerebral blood flow.

9.3 The Role of Nitrates in Cerebrovascular Disease

Cerebrovascular disease is recognized as a leading cause of death globally, with a prevalence of up to 25% in adults above 25 in recent years. China exhibits the highest risk of stroke, which is the primary cause of disability in the country [13]. Stroke and its complications impose significant burdens on patients, families, and society. Thrombolysis and endovascular thrombus removal are the only available treatments for acute ischemic stroke; however, only a small proportion of ischemic stroke patients can benefit from these treatments due to a limited treatment time window [14]. Research indicates that precursor pathophysiological changes, such as endothelial dysfunction, can be detected prior to morphological changes in the cerebral vascular wall, subsequently influencing multiple pathways involving smooth muscle cells and blood cells, etc. [15].

9.3.1 Nitrate Metabolism in Ischemic Cerebrovascular Disease

Nitrate metabolism plays a crucial role in the development of ischemic cerebrovascular disease. Studies have shown that chronic dietary nitrate deficiency in mice can lead to endothelial dysfunction, and cerebral tissue ischemia and hypoxia can intensify NO production in a compensatory way. Some researchers conducted experiments on a rat model of total cerebral ischemia and found that the protective properties of nitrates during cerebral hypoxia-ischemia and NOS inhibition depended on the cation and the concentration of nitrates [16]. It has also been found that the mechanism of nitrate/nitrite and NO in ischemic diseases may be related to the mitochondrial respiratory chain and oxidative stress [2]. Nitrites exert an inhibitory effect on the electron transport chain through mercapto-nitrosation and reduce the production of reactive oxygen species after reperfusion. It is speculated that nitrites may be involved in the improvement of cerebrovascular ischemia-reperfusion.

Meanwhile, some scholars have found that NO can directly mediate vasodilation by triggering dephosphorylation of myosin light chain through the cGMP pathway and by increasing the opening rate of large-conductance calcium-activated potassium channels [17] (Fig. 9.2). In addition, NO can also affect the synthesis of prostaglandins and other mediators to indirectly regulate vasodilation, regulate the above processes, or promote new blood vessels formulation, remodeling small venous vessels to regulate cerebral angiogenesis and improve the prognosis of neurological function after stroke [18]. It has been found that the dose of nitrites required to exert a protective effect on the cerebrovascular endothelium is low in ischemia-reperfusion injury models, whereas high doses of nitrites may produce

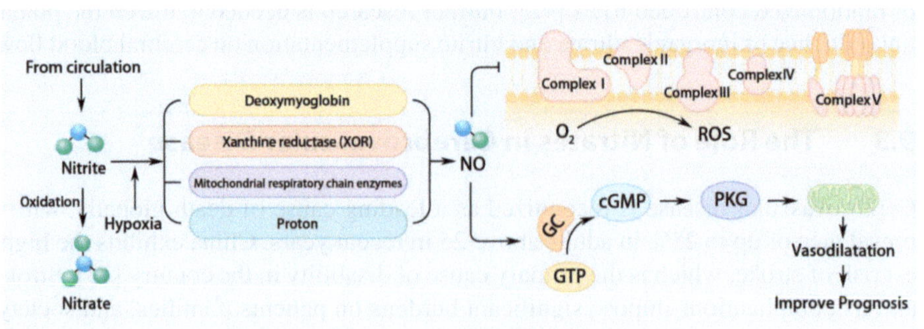

Fig. 9.2 The mechanisms of nitrate/nitrite and NO in ischemic hypoxic diseases

neurotoxic effects through oxidation of proteins and lipids, causing further damage to brain tissue. Therefore, the role of NO or nitrate/nitrite products in ischemic stroke depends mainly on their source, local concentration, and redox state, etc. [19].

The damage-associated molecular pattern (DAMP) generated by ischemia and hypoxia following cerebral ischemia can swiftly activate the corresponding immune response. Consequently, NO produced by this immune response exerts toxic effects on adjacent tissues and on cells capable of synthesizing NOS. Specific local or systemic tissue injuries associated with the immune system, as well as abnormal dilation and permeability of blood and lymphatic vessels, may be closely linked to the local concentration of NO.

9.3.2 Nitrates Protection Against Focal Ischemic Stroke in Mice

Songlin Wang's group practiced oral inorganic nitrate pretreatment in a mouse model of focal cerebral ischemia, using multiple drug concentrations, and found that nitrate concentration is highly relevant to the recovery of neurological function in mice after surgery. The high-dose group would have a strong neuroprotective effect in the early stage, while the chronic stage of the disease would show some slowing down of neurological function recovery and so on. The low-dose group showed a continuous and stable trend toward better recovery. We found that nitrate-treated ischemic mice showed a significant reduction in neuronal apoptosis and infarct volume in the early stage, which is consistent with scholars' use of NO to regulate blood flow to improve ischemic stroke. Later, we observed that pretreatment with nitrates still had positive therapeutic significance on mice stroke by cutting off the NO production pathway by removing the intestinal flora, suggesting that other mechanisms may be involved in the therapeutic mechanism of nitrate pretreatment on ischemic stroke in mice besides NO.

In order to verify this possible other mechanism, Songlin Wang's group explored the role of nitrates by performing in vitro cell cultures. Since in vitro cell culture is performed in a sterile environment, the pathway of NO production from nitrates was thought to be absent in vitro. Interestingly, by subjecting primary neurons to hypoxia, it was found that cell survival increased more than onefold in the low-dose nitrate group. Nitrates promote proliferation and state maintenance of neural stem cells and their differentiation toward neurons. All these indicate that the exclusion of the NO pathway in vivo and in vitro could still prevent the progression of stroke, suggesting the existence of a potential new mechanism. At this time, Songlin Wang's group discovered that nitrate's improved functional recovery after stroke may be related to the nitrate transporter protein, sialin. Neural stem cells almost all expressed sialin, and the expression of sialin was higher in nitrate-treated neural stem cells (Fig. 9.3). These illustrate the potential of nitrate/nitrite-No/sialin in ischemic stroke therapy.

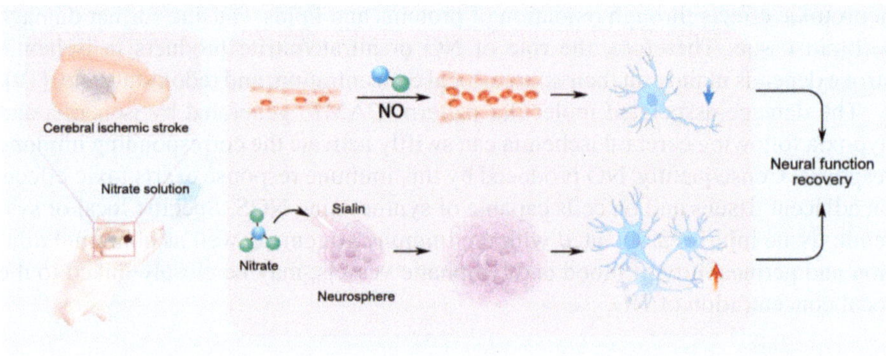

Fig. 9.3 Mechanism of action of nitrates in the treatment of ischemic stroke in mice

9.3.3 Nitrates and Associated Cerebrovascular Disorders

Cerebrovascular diseases, including cerebral hemorrhage and small vessel disease, are associated with cerebral vascular lesions and oxidative stress. Research indicates that cerebral vasospasm following cerebral hemorrhage correlates with decreased NO levels in the acute phase [20]. Additionally, this condition may be associated with eNOS dysfunction [21]. Clinical findings reveal that patients post-cerebral hemorrhage exhibit elevated nitrate concentrations in cerebrospinal fluid compared to healthy controls, while plasma nitrate levels are diminished. Nitrite infusion for these patients has been shown to significantly alleviate vasospasm.

9.4 The Role of Nitrates in Neurodegenerative Diseases

Neuronal loss is significantly linked to the pathological progression of neurodegenerative diseases, including Alzheimer's disease and vascular dementia (VD). The degeneration of neuronal and synaptic functions occurs progressively due to various risk factors, resulting in impaired brain function. Furthermore, cerebrovascular diseases, such as ischemic stroke, hemorrhagic cerebrovascular disease, or cerebral small vessel disease, are commonly observed in the elderly population. These conditions, characterized by vasospasm, stenosis, or hemorrhage of blood vessels, lead to a reduction in cerebral blood flow (CBF). Decreased CBF can result in diminished oxygen and nutrient supply to the brain, further exacerbating the decline in memory and cognitive function.

NO supplements, such as nitroglycerin, have been reported to significantly reverse excitotoxicity both in vitro and in vivo. NO-supplemented drugs, including nitroglycerin and sodium nitroprusside, are clinically available for the treatment of cardiovascular disease. However, the poor distribution and short half-life of these drugs in the brain limit their application in the treatment of neurodegenerative diseases. Meanwhile, long-term use of NO-supplemented drugs may carry a risk of hypotension, as systemic release of NO may lead to peripheral vasodilation. Therefore, an ideal NO-supplemented drug for the treatment of vascular dementia should be able to release NO specifically in brain tissue to avoid the risk of hypotension. In order to achieve dual functions such as neuroprotection and vasodilation, meperidine nitrate [22] has been designed and synthesized. With a unique mechanism of action, memantine is the only N-methyl-D-aspartate receptor (NMDAR) antagonist, which alleviates neuronal loss by inhibiting over-activation after glutamate binding and reducing excessive calcium inward flow. Memantine amine nitrate is dosed with the addition of nitrates to memantine. Memantine nitrate has a concentration-dependent protective effect against glutamate-induced neurotoxicity in primary cultured rat cerebellar granule neurons (CGNs). It may be neuroprotective through inhibition of the ERK pathway and concomitant activation of the PI3K/Akt pathway. In addition, depending on its concentration, it can dilate preconstricted middle cerebral arteries and improve cerebral blood flow by activating the NO pathway in vitro (Fig. 9.4). In a rat model of vascular occlusion, memantine nitrate may attenuate spatial memory impairments and motor dysfunction through neuroprotection and improved cerebral blood flow. The combination of nitrate and memantine may have further implications for the prevention and treatment of neurodegenerative diseases.

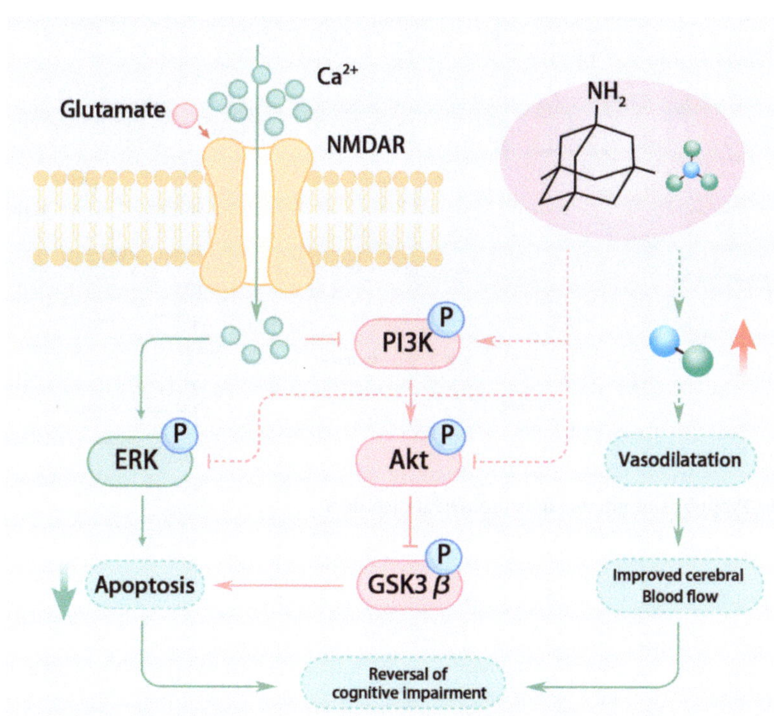

Fig. 9.4 Mechanism of memantine nitrate in cognitive protection

9.5 Prospects for Nitrate Utilization in Neurological Disorders

In the future, administration of exogenous nitrates/nitrites or other precursors of NO may become a new clinical treatment option for the treatment of cerebrovascular and neurodegenerative diseases, but a large number of basic and clinical studies are still needed to fortify the evidence. Although current studies have found that supplementation of nitrates and sialates by drugs or diet or even other methods may be beneficial in improving lesions, their dose and effects in the body are affected by the daily administration of proton pump inhibitors, related drugs that inhibit xanthine oxidoreductase, and antimicrobial mouthwashes [23]. However, the manner and dosage of nitrates for different neurological disorders need to be further explored. In addition, the heterogeneity of nitrates and nitrites in different patients deserves further study [24]. At the current stage, it may be possible to measure nitrates/nitrites in blood, urine, and saliva as an early warning molecule of disease for the purpose of timely prevention. In conclusion, nitrates have positive significance in disease protection of the nervous system, and more basic experimental data support is needed to advance clinical application (Fig. 9.5).

Compliance with Ethical Standards All applicable international, national, and/or institutional guidelines for the care and use of animals were followed.

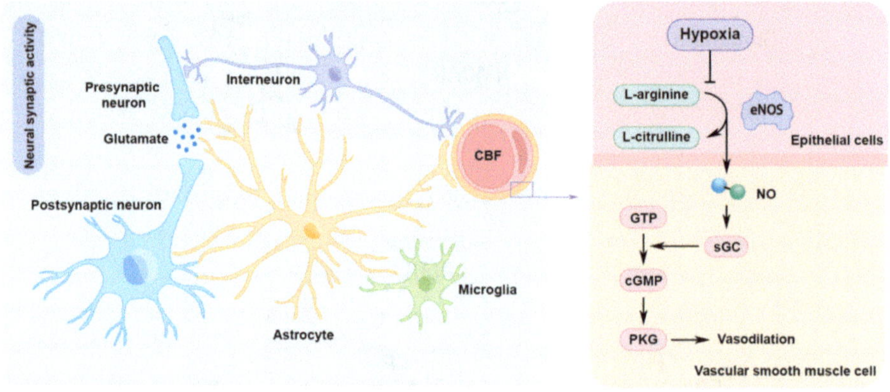

Fig. 9.5 Prognosis of nitrates in neurological disorders

References

1. Feng Y, Cao X, Zhao B, et al. Nitrate increases cisplatin chemosensitivity of oral squamous cell carcinoma via REDD1/AKT signaling pathway. Sci China Life Sci. 2021;64(11):1814–28.
2. Shiva S, Sack MN, Greer JJ, et al. Nitrite augments tolerance to ischemia/reperfusion injury via the modulation of mitochondrial electron transfer. J Exp Med. 2007;204(9):2089–102.
3. Furchgott RF, Bhadrakom S. Reactions of strips of rabbit aorta to epinephrine, isopropylarterenol, sodium nitrite and other drugs. J Pharmacol Exp Ther. 1953;108(2):129–43.
4. Palmer RM, Ferrige AG, Moncada S. Nitric oxide release accounts for the biological activity of endothelium-derived relaxing factor. Nature. 1987;327(6122):524–6.
5. Murad F. Shattuck Lecture. nitric oxide and cyclic GMP in cell signaling and drug development. N Engl J Med. 2006;355(19):2003–11.
6. Fukuda T, Kakinohana M, Takayama C, et al. Dietary supplementation with sodium nitrite can exert neuroprotective effects on global cerebral ischemia/ reperfusion in mice. J Anesth. 2015;29(4):609–17.
7. Ben Achour S, Pascual O. Glia: the many ways to modulate synaptic plasticity. Neurochem Int. 2010;57(4):440–5.
8. Jansson EA, Huang LY, Malkey R, et al. A mammalian functional nitrate reductase that regulates nitrite and nitric oxide homeostasis. Nat Chem Biol. 2008;4(7):411–7.
9. Lundberg JO, Weitzberg E, Gladwin MT. The nitrate-nitrite-nitric oxide pathway in physiology and therapeutics. Nat Rev Drug Discov. 2008;7(2):156–67.
10. Qin LZ, Liu XB, Sun QF, et al. Sialin (SLC17A5) functions as a nitrate transporter in the plasma membrane. Proc Natl Acad Sci U S A. 2012;109(33):13434–9.
11. Tabatabaei SN, Girouard H. Nitric oxide and cerebrovascular regulation. Vitam Horm. 2014;96:347–85.
12. Clifford T, Babateen A, Shannon OM, et al. Effects of inorganic nitrate and nitrite consumption on cognitive function and cerebral blood flow: a systematic review and meta-analysis of randomised clinical trials. Crit Rev Food Sci Nutr. 2019;59(15):2400–10.
13. GBD 2016 Lifetime Risk of Stroke Collaborators, Feigin VL, Nguyen G, et al. Global, regional, and country-specific lifetime risks of stroke, 1990 and 2016. N Engl J Med. 2018;379(25):2429–37.
14. Meschia JF, Brott T. Ischaemic stroke. Eur J Neurol. 2018;25(1):35–40.
15. Poggesi A, Pasi M, Pescini F, et al. Circulating biologic markers of endothelial dysfunction in cerebral small vessel disease: a review. J Cereb Blood Flow Metab. 2016;36(1):72–94.
16. Kuzenkov VS. Protective role of nitrate/nitrite reductase system during transient global cerebral ischemia. Bull Exp Biol Med. 2018;165(1):31–5.
17. Dormanns K, Brown RG, David T. The role of nitric oxide in neurovascular coupling. J Theor Biol. 2016;394:1–17.
18. Lapi D, Colantuoni A. Remodeling of cerebral microcirculation after ischemia-reperfusion. J Vasc Res. 2015;52(1):22–31.
19. Rashid PA, Whitehurst A, Lawson N, et al. Plasma nitric oxide (nitrate/nitrite) levels in acute stroke and their relationship with severity and outcome. J Stroke Cerebrovasc Dis. 2003;12(2):82–7.
20. Sakowitz OW, Wolfrum S, Sarrafzadeh AS, et al. Relation of cerebral energy metabolism and extracellular nitrite and nitrate concentrations in patients after aneurysmal subarachnoid hemorrhage. J Cereb Blood Flow Metab. 2001;21(9):1067–76.
21. Sabri M, Ai J, Knight B, et al. Uncoupling of endothelial nitric oxide synthase after experimental subarachnoid hemorrhage. J Cereb Blood Flow Metab. 2011;31(1):190–9.
22. Mak S, Liu Z, Wu LM, et al. Pharmacological characterizations of anti-dementia memantine nitrate via neuroprotection and vasodilation *in vitro* and *in vivo*. ACS Chem Neurosci. 2020;11(3):314–27.

23. Qin LZ, Jin LY, Qu XM, et al. Nitrate: a pioneer from the mouth to the systemic health and diseases. Science. 2020;55(7):433–8.
24. Smith EE, Saposnik G, Biessels GJ, et al. Prevention of stroke in patients with silent cerebrovascular disease: a scientific statement for healthcare professionals from the American Heart Association/American Stroke Association. Stroke. 2017;48(2):e44–71.

Metabolic Diseases and Aging Protection

10

Yipu Xu, Xiaoyu Li, Linsha Ma, Yifan Xu, and Jian Zhou

Aging is associated with a gradual decline in metabolism, leading to the prevalence of metabolic diseases such as osteoporosis and obesity. The relationship between metabolic diseases and aging is closely linked to the dysfunction of various signaling molecules in the body. NO, a critical signaling molecule, is essential for glucose, lipid, and energy metabolism and also contributes to aging protection. Nitrate-nitrite serves as a significant means of maintaining NO levels in the body. Our findings indicate that exogenous nitrate supplementation can effectively prevent osteoporosis and obesity, slow the aging process, and play a crucial role in preventing metabolic diseases associated with aging.

10.1 Osteoporosis Prevention

Yipu Xu

10.1.1 Definition and Pathogenesis of Osteoporosis

Osteoporosis is a prevalent metabolic bone disease characterized by osteopenia and bone degeneration. It predominantly affects postmenopausal women and the elderly, leading to hip fractures and long bone fractures, significantly impairing life quality [1]. Research indicates that mesenchymal stem cell function tends to be hindered in

Y. Xu · X. Li · Y. Xu · J. Zhou (✉)
Beijing Stomatological Hospital, Capital Medical University, Beijing, China
e-mail: xiaoyu9449@ccmu.edu.cn; zhoujian@ccmu.edu.cn

L. Ma
Beijing Friendship Hospital, Capital Medical University, Beijing, China

S. Wang (ed.), *Inorganic Nitrate and Human Body Homeostasis*,
https://doi.org/10.1007/978-981-95-2824-0_10

osteoporosis, manifesting in several ways: weakened osteogenic capacity of mesenchymal stem cells, intensified lipid formation capacity, and a decline in immunomodulatory properties. Pharmacological agents that restore stem cell function can effectively mitigate osteoporosis symptoms [2].

10.1.2 Nitrates Contribute to the Prevention of Osteoporosis

In 2003, a retrospective study indicated that patients with cardiovascular disease who used organic nitrates for a long time exhibited a lower incidence and severity of osteoporosis compared to those who did not use nitrates. Nonetheless, the limitations of organic nitrates, including their short duration of action, significant side effects, and inconsistent efficacy, render them unsuitable as a therapeutic option for osteoporosis [3].

In an animal experiment, administration of 2 mmol/L nitrate to osteoporosis model rats effectively prevented bone loss symptoms. Improvements were observed in bone morphology indices, including bone volume fraction (BV/TV), trabecular spacing (Tb.Sp), and bone density (BMD). Additionally, relevant indices of bone metabolism were enhanced (Fig. 10.1). Nitrates also demonstrated a therapeutic effect in early osteoporosis.

10.1.3 Mechanism of Action by Which Nitrates Prevent Osteoporosis

Bone mesenchymal stem cells (BMMSCs) are derived from bone marrow and possess significant multidirectional differentiation potential and self-renewal capabilities. They can differentiate into both osteogenic and lipogenic lineages. Additionally,

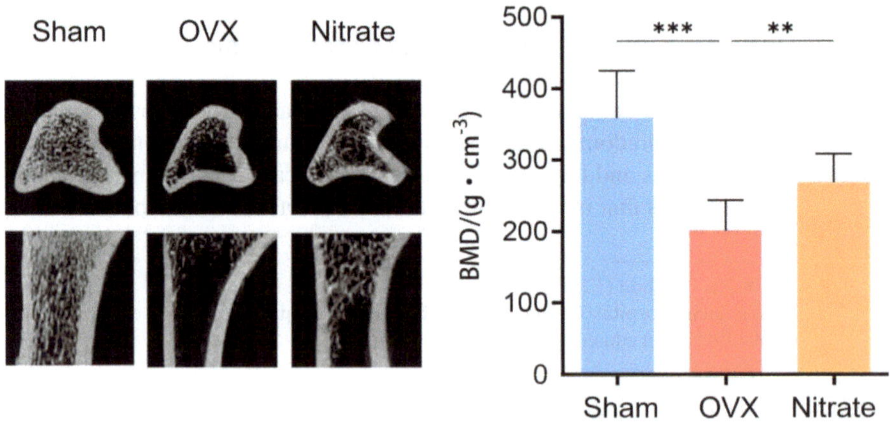

Fig. 10.1 Oral nitrates can effectively prevent the bone loss in osteoporosis rat

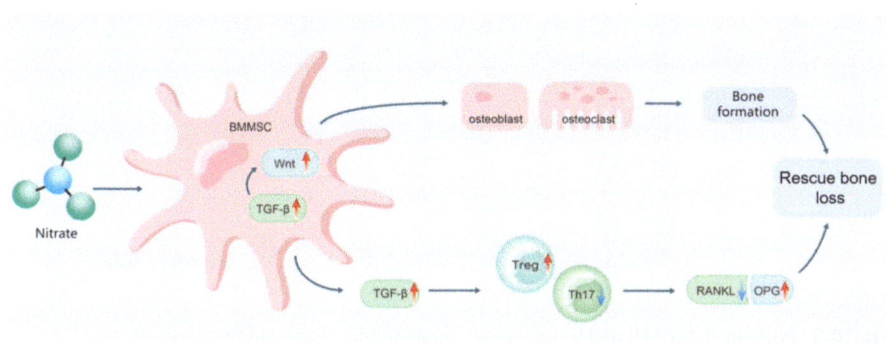

Fig. 10.2 The mechanism by which oral nitrates prevent bone loss in a rat model of osteoporosis

BMMSCs secrete various cytokines and play a crucial role in immune regulation. In cases of osteoporosis caused by estrogen deficiency and aging, the differentiation ability of BMMSCs exhibits varying degrees of impairment [4]. NO is a vital molecule for maintaining stem cells; thus, supplementation with nitrates through the nitrate-nitrite-NO pathway may enhance the osteogenic capacity of stem cells, inhibit abnormal lipid formation, and restore the stem cells' immunoregulatory functions.

Nitrates may enhance TGF-β expression in stem cells and activate the Wnt pathway associated with osteogenesis through NO. They can inhibit the increase in Th17 cell proportions due to abnormal cell death by secreting TGF-β, improve Treg cell proportions, balance osteoclast and osteoblast cytokines via RANKL/OPG, reduce bone destruction, and promote bone formation, thereby alleviating osteoporosis symptoms (Fig. 10.2).

10.2 Aging Protection

Xiaoyu Li

According to the seventh population census in 2020, China's population aged 60 and above is 264 million, representing 18.70% of the total population. This figure significantly exceeds the threshold of 10% that defines an aging society, indicating that China has stepped over the threshold. As the proportion of the aging population increases and the aging society deepens, the management of aging-related diseases has emerged as a prominent and challenging area of research. Studies indicate that nitrate levels decline with age, and plasma nitrite concentrations in older individuals are lower after nitrate consumption compared to younger participants, which confirms a reduction in exogenous NO bioavailability in the elderly [5]. Aging is frequently linked to diminished NO bioactivity, which is associated with various age-related diseases.

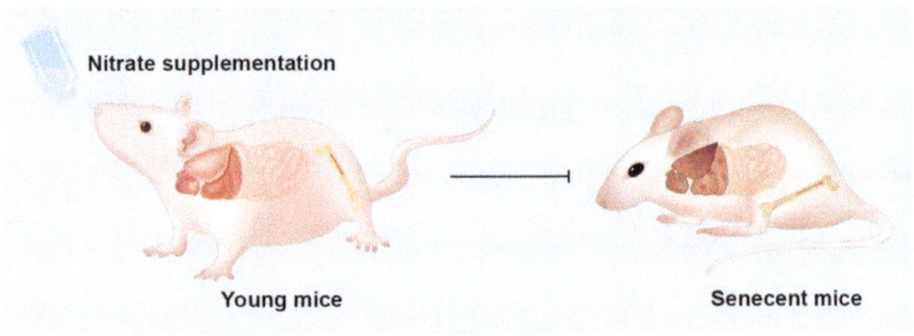

Fig. 10.3 Nitrates delay cell and tissue aging and maintain functional homeostasis

Metabolic syndrome, endothelial dysfunction, and cardiovascular death were observed in a low nitrite/nitrate diet mouse model [6]. This suggests that nitrate deficiency may accelerate aging and lead to cardiovascular death. Songlin Wang's group observed decreased nitrate levels and liver degeneration in β-galactosidase-induced senescent mice and naturally senescent mice. Besides, daily nitrate supplementation significantly restored plasma nitrate levels, reduced glutamate aminotransferase and aspartate aminotransferase levels, and prevented hepatic tissue cellular senescence and degradation of the structure and glycolipid metabolism [7]. This suggests that nitrates have the potential to prevent aging-related hepatic degeneration. In addition, Songlin Wang's group found that dietary nitrates can prevent osteoporosis in the femur of naturally aging mice and delay the aging of bone marrow mesenchymal stem cells. Meanwhile, *in vitro* nitrate supplementation slowed etoposide- and hydrogen peroxide (H_2O_2)-induced senescence of MSCs (Fig. 10.3).

10.3 Obesity Prevention

Linsha Ma

10.3.1 Global Obesity Trends and Their Significant Health Implications

Obesity is defined as an abnormal or excessive accumulation of body fat that poses a risk to health. Epidemiological data provided by the World Health Organization (WHO) indicates that between 1975 and 2016, the proportion of obesity nearly tripled. In 2016, 39% of individuals over the age of 18 were overweight (over 1.9 billion people), with 39% of males and 40% of females being overweight; approximately 13% were obese (655 million). Obesity, once thought to be more prevalent in high-income countries, is now also showing a clear upward trend in middle- and low-income countries, especially in urban areas [8]. Obesity is a multifactorial metabolic imbalance that poses a threat to health and is a major risk factor for a range of chronic diseases [11], including diabetes, cardiovascular diseases, inflammatory diseases, and cancer. The risk of these non-communicable diseases is significantly associated with Body Mass Index (BMI). According to the 2017 Global Burden of Disease statistics, over 4 million deaths per year are attributed to overweight or obesity [9].

10.3.2 Strategies for Controlling Obesity

The mechanism underlying obesity is fundamentally characterized by an imbalance between energy intake and consumption, which includes an increase in the consumption of high-energy foods and a decrease in physical activity. Furthermore, disruptions in fat metabolism and imbalances in gut microbiota are significant contributors to obesity. Physiologists believe that obesity and its associated diseases are largely preventable, prompting investigations into effective prevention and treatment strategies that target body metabolism regulation and gut microbiota adjustment [10–12]. Dietary regulation for weight control is regarded as the safest approach and remains a prominent area of research.

10.3.3 Regulation of Gut Microbiota by Nitrates to Prevent Obesity

In metabolic diseases, nitrates primarily function through the NO-cGMP pathway, including regulating NADPH to reduce oxidative stress and protect tissue, as well as promoting adenosine monophosphate (AMP)-activated protein kinase and downstream pathways, which regulate fat synthesis, fatty acid oxidation, and blood sugar homeostasis [13–15]. Songlin Wang's research group, through the establishment of a high-fat diet mouse model, found that adding nitrates to drinking water can alleviate obesity, reducing fat weight and adipocyte size, improving glucose tolerance and insulin sensitivity, and alleviating hyperlipidemia and hepatic fat deposition. This finding suggests that nitrates have an improving effect on metabolism and a preventive role against diet-induced obesity. Moreover, nitrates can regulate the gut microbiota in high-fat diet mice. 16s rRNA sequencing and bio-informatics analysis found that a high-fat diet significantly alters the diversity and composition of the gut microbiota in mice. High-fat diet mice had fewer operational taxonomic units (OTUs), lower microbial abundance, and reduced microbial diversity. Principal component analysis (PCA) suggested that the gut microbiota of mice which drank nitrate-supplemented water tended to move closer, indicating that nitrates can partially alleviate the gut microbiota disorder caused by a high-fat diet. Through differential analysis of gut microbiota, beneficial bacteria such as lactobacilli and bifidobacteria were significantly increased in the nitrate-supplemented group, while Bacteroides, which are positively correlated with obesity, were also significantly decreased [16]. In addition, other studies by Songlin Wang's group on the regulation of gut microbiota by nitrates implied that oral nitrates could alleviate DSS-induced intestinal inflammation in mice, up-regulating the abundance of beneficial bacteria such as lactobacilli, ruminococcaceae, and prevotella, and down-regulating the abundance of pathogens such as Bacteroidales_S24-7_group_unidentified and Bacteroides [17]. After whole-body radiation in rats, nitrates relieved colitis by regulating intestinal microbial homeostasis, with a significant increase in lactobacilli in the gut [18].

10.4 Hypoxia Protection

Yifan Xu and Jian Zhou

Oxygen is essential for sustaining metabolic processes in organisms and is fundamental to life activities. A significant reduction in environmental oxygen levels triggers various responses in critical systems and organs, including cerebral edema, pulmonary edema, developmental disorders of the central nervous system, atherosclerosis, and osteoporosis, all of which impair normal development and metabolism. Therefore, strategies to safeguard organisms from hypoxia-related damage have become a pivotal area of research.

Plateaus represent common hypoxic environments. At sea level, atmospheric pressure is 760 mmHg, with an oxygen content of 21%. As altitude increases, air pressure decreases, resulting in thinner air and reduced oxygen content. At an average altitude of 3650 m in Lhasa, atmospheric pressure falls to 489 mmHg, and oxygen content decreases to 15%. In such oxygen-deficient conditions, organisms experience rapid physiological responses in brain and lungs. Typically, gradual acclimatization to plateau regions may lead to symptoms such as delayed reactions, impaired memory, and hallucinations. Conversely, rapid ascent can result in insomnia, dizziness, headaches, or acute altitude sickness, including loss of consciousness and respiratory difficulties, significantly impacting the health (Fig. 10.4).

NO is important for preventing hypoxia injury. Many scholars contend NO to be a big part of the acclimatization process at high altitudes [19]. Nitrate is a major source of NO in the body and helps to keep balance and protect against hypoxia. In 2011, Levett [20] compared the plasma nitrate levels of mountaineers climbing Mount Everest and found that as they climbed higher, their nitrate levels went up significantly, sometimes reaching up to 10 times the level found at sea level. More recently, Italian and Canadian researchers did a similar study [21]. They tested the nitrate levels in the blood of 15 mountaineers living in the plains before and after climbing Mount Rosa (4554 m above sea level) and Mount Solda (3269 m above sea level). They found that on the first day after entering the plateau, the nitrate concentration in all the volunteers' blood increased significantly. The higher the altitude, the longer it took for the plasma nitrate ion concentration to reach its peak (Fig. 10.5) [22]. This suggests that nitrates might help the body adjust to low-oxygen levels when entering a hypoxic environment. The research group had already drunk nitrate-rich beet juice to prevent altitude sickness because when you take in extra

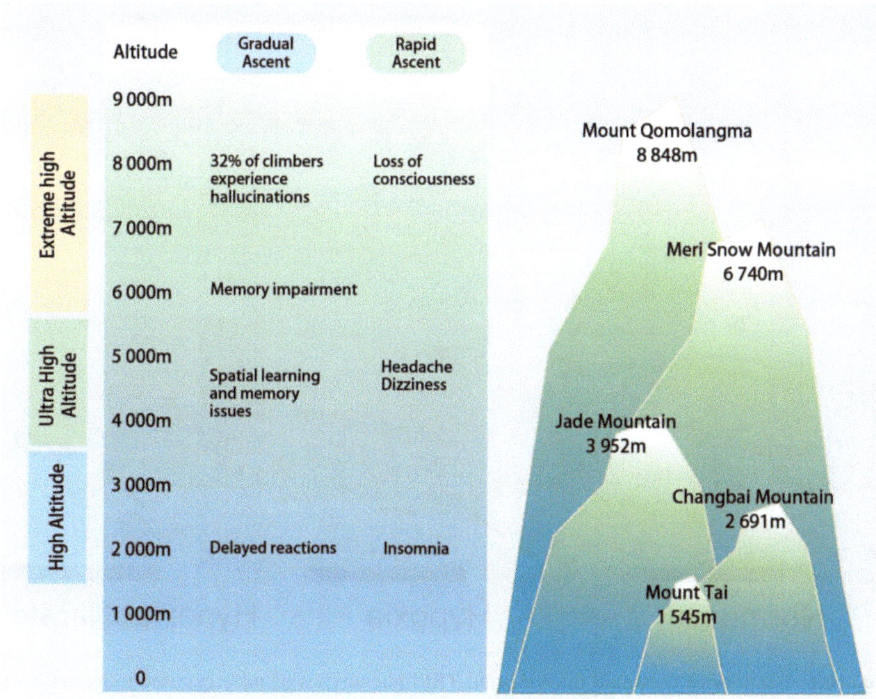

Fig. 10.4 Altitude sickness manifests at varying elevations

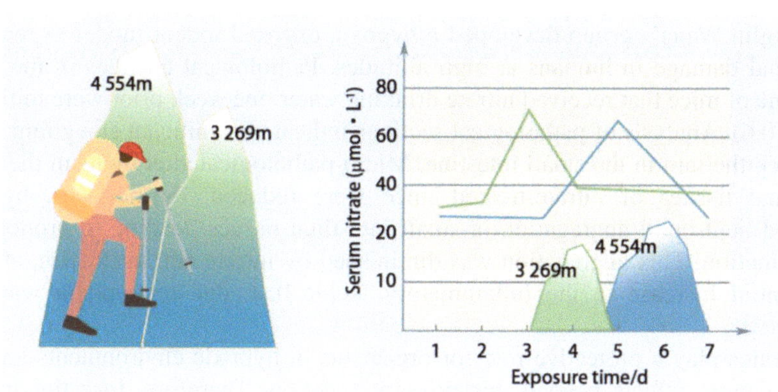

Fig. 10.5 Plasma nitrates trend following climbers' entry onto the plateau

nitrates, it can help athletes and climbers adjust to the plateau and improve their performance.

Gastrointestinal discomfort in hypoxic environments, while less severe than acute cerebral and pulmonary edema, presents symptoms such as loss of appetite,

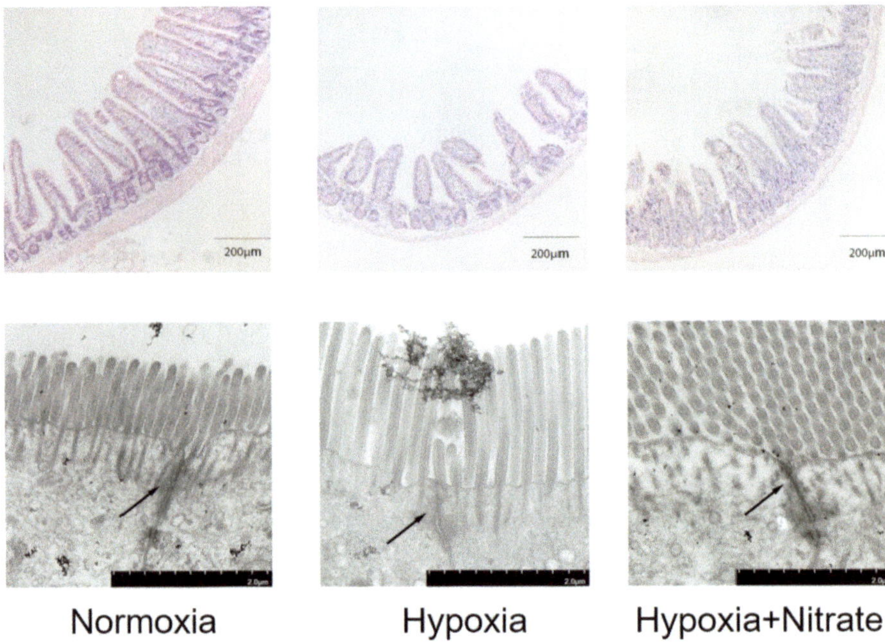

Fig. 10.6 Sections of the small intestine and TEM phenotypes of mice exposed to hypoxia and treated with normal oxygen, hypoxia, and nitrate

flatulence, and diarrhea. All of these symptoms can significantly impact individuals' daily lives and work upon entering the plateau.

Songlin Wang's group developed a hypoxia-exposed rodent model to replicate intestinal damage in humans at high altitudes. Pathological changes in the small intestine of mice that received nitrate drinking water one week prior were mitigated (Fig. 10.6). Analysis of pathological sections indicated significant sloughing of the villus epithelium in the small intestine. Micro-pathological alterations in the small intestinal tissues of nitrate-treated mice were reduced. Furthermore, hypoxia induced neutrophil aggregation in small intestinal tissues, leading to pronounced inflammation. This aggregation was diminished by nitrate administration, and the substantial increase in the inflammatory factor IL-6 due to hypoxia was also alleviated.

Nitrates play a protective role for organisms in hypoxic environments and can achieve more effects beyond this primary function. Therefore, investigating the mechanisms by which nitrates operate in hypoxic conditions and identifying additional hypoxia-related diseases that it may mitigate will be the focus of future research.

Compliance with Ethical Standards All applicable international, national, and/or institutional guidelines for the care and use of animals were followed.

References

1. Cauley JA. Public health impact of osteoporosis. J Gerontol A Biol Sci Med Sci. 2013;68(10):1243–51.
2. Demontiero O, Vidal C, Duque G. Aging and bone loss: new insights for the clinician. Ther Adv Musculoskelet Dis. 2012;4(2):61–76.
3. Misra D, Peloquin C, Kiel DP, et al. Intermittent nitrate use and risk of hip fracture. Am J Med. 2017;130(2):229.e15–20.
4. Qi M, Zhang LQ, Yang M, et al. Autophagy maintains the function of bone marrow mesenchymal stem cells to prevent estrogen deficiency-induced osteoporosis. Theranostics. 2017;7(18):4498–516.
5. Toprakçi M, Ozmen D, Mutaf I, et al. Age-associated changes in nitric oxide metabolites nitrite and nitrate. Int J Clin Lab Res. 2000;30(2):83–5.
6. Kina-Tanada M, Sakanashi M, Tanimoto A, et al. Long-term dietary nitrite and nitrate deficiency causes metabolic syndrome, endothelial dysfunction, and cardiovascular death in mice. Diabetologia. 2017;60(6):1138–51.
7. Wang HF, Hu L, Li L, et al. Inorganic nitrate alleviates the senescence-related decline in liver function. Sci China Life Sci. 2018;61(1):24–34.
8. NCD Risk Factor Collaboration (NCD-RISC. Worldwide trends in body-mass index, underweight, overweight, and obesity from 1975 to 2016: a pooled analysis of 2416 population-based measurement studies in 128.9 million children, adolescents, and adults. Lancet. 2017;390(10113):2627–42.
9. Flegal KM, Kit BK, Orpana H, et al. Association of all-cause mortality with overweight and obesity using standard body mass index categories: a systematic review and meta-analysis. JAMA. 2013;309(1):71–82.
10. Cano PG, Santacruz A, Trejo FM, et al. Bifidobacterium CECT 7765 improves metabolic and immunological alterations associated with obesity in high-fat diet-fed mice. Obesity (Silver Spring). 2013;21(11):2310–21.
11. Lee E, Jung SR, Sy L, et al. *Lactobacillus plantarum* strain Ln4 attenuates diet-induced obesity, insulin resistance, and changes in hepatic mRNA levels associated with glucose and lipid metabolism. Nutrients. 2018;10(5):643.
12. Long RT, Zeng WS, Chen LY, et al. Bifidobacterium as an oral delivery carrier of oxyntomodulin for obesity therapy: inhibitory effects on food intake and body weight in overweight mice. Int J Obes (Lond). 2010;34(4):712–9.
13. Roberts LD, Ashmore T, Kotwica AO, et al. Inorganic nitrate promotes the browning of white adipose tissue through the nitrate-nitrite-nitric oxide pathway. Diabetes. 2015;64(2):471–84.
14. Roberts LD. Does inorganic nitrate say NO to obesity by browning white adipose tissue? Adipocyte. 2015;4(4):311–4.
15. Cordero-Herrera I, Kozyra M, Zhuge Z, et al. AMP-activated protein kinase activation and NADPH oxidase inhibition by inorganic nitrate and nitrite prevent liver steatosis. Proc Natl Acad U S A. 2019;116(1):217–26.
16. Ma LS, Hu L, Jin L, et al. Rebalancing glucolipid metabolism and gut microbiome dysbiosis by nitrate-dependent alleviation of high-fat diet-induced obesity. BMJ Open Diabetes Res Care. 2020;8(1):e001255.
17. Hu L, Jin LY, Xia DS, et al. Nitrate ameliorates dextran sodium sulfate-induced colitis by regulating the homeostasis of the intestinal microbiota. Free Radic Biol Med. 2020;152:609–21.
18. Wang WL, Hu L, Chang SM, et al. Total body irradiation-induced colon damage is prevented by nitrate-mediated suppression of oxidative stress and homeostasis of the gut microbiome. Nitric Oxide. 2020;102:1–11.
19. Umbrello M, Dyson A, Feelisch M. The key role of nitric oxide in hypoxia: hypoxic vasodilation and energy supply-demand matching. Antioxid Redox Signal. 2013;19(14):1690–710.
20. Levett DZ, Femrnandez BO, Riley HL, et al. Caudwell extreme Everest research group. The role of nitrogen oxides in human adaptation to hypoxia. Sci Rep. 2011;1:109.

21. Rasica L, Porcelli S, Limper U, et al. Beet on Alps: Time-course changes of plasma nitrate and nitrite concentrations during acclimatization to high-altitude. Nitric Oxide. 2021;107:66–72.
22. Xu Y, Sa Y, Zhang C, et al. A preventative role of nitrate for hypoxia-induced intestinal injury. Free Radic Biol Med. 2024;213:457–69.

Part IV

The Clinical Application of Nitrate

As our understanding of the relationship between nitrates and human health has improved, nutritional supplements containing nitrates have been developed. The beneficial effects of these supplements on the human body have been confirmed by an increasing number of clinical studies. Maximizing the efficacy of oral nitrate supplements and ensuring that they are safe and controllable are critical issues in determining whether they can be applied to further improve human health. To address this issue, a new type of nitrate drug has been developed.

Efficacy and Limitations of Nutritional Supplements Containing Nitrates as Their Main Component

11

Xingmin Qu

The body's nitrate sources include both exogenous nitrates from food and endogenous nitrates formed via the oxidation of nitrites and nitric oxide (NO). Exogenous nitrates can be transformed into nitrites by oral bacteria through the digestive tract-salivary gland cycle and are eventually converted into NO in the blood and tissues. Therefore, they are considered important sources of NO in the body. The currently employed methods for loading exogenous nitrates include oral administration of potassium nitrate or sodium nitrate, as well as the ingestion of natural sources such as vegetable juices. In clinical studies, compared with potassium nitrate and sodium nitrate, vegetable juices rich in nitrates, most commonly beetroot juice (BRJ), have been used more widely as nutritional supplements. To meet the needs of clinical studies, manufacturers may isolate nitrates from vegetable juices to serve as a negative control. In this book, natural vegetable juice is presented as a nutritional supplement with nitrates as its main component.

11.1 Efficacy of Nutritional Supplements with Nitrates as Their Main Component

In clinical studies where nutritional supplements with nitrates as their main component are used as interventions, the efficacy of nitrates in the human body are mainly demonstrated through improvements in both cardiovascular function and exercise capacity.

X. Qu (✉)
Beijing Stomatological Hospital, Capital Medical University, Beijing, China

S. Wang (ed.), *Inorganic Nitrate and Human Body Homeostasis*,
https://doi.org/10.1007/978-981-95-2824-0_11

11.1.1 Cardiovascular Function

As early as 2008, researchers discovered that consuming a single bottle of nitrate-rich beetroot juice (500 mL, containing approximately 1400 mg of nitrates) could reduce systolic blood pressure by 10 mmHg and diastolic blood pressure by 8 mmHg [1]. This antihypertensive effect was similar to that of an antihypertensive drug. The peak antihypertensive effect was consistent with the plasma nitrite content and could last for 24 h. Even a lower dose of beetroot juice (250 mL, containing approximately 340 mg of nitrates) could reduce systolic blood pressure by 5 mmHg [2]. Additionally, the observed increase in plasma cyclic guanosine monophosphate (cGMP) levels indicated that the NO produced by the reduction of nitrite activates the soluble guanylate cyclase (sGC)-cGMP-mediated vasodilation [1]. *In vitro* studies have shown that beetroot juice can delay platelet aggregation [1, 3]. Another study confirmed the antihypertensive effect of dietary nitrates [4], which was generally dependent on plasma nitrite levels. Moreover, some researchers found that the dose-dependent on systolic blood pressure effect was more pronounced when investigating the antihypertensive effect of three doses of beetroot juice [5].

Most of the above studies focused on healthy individuals, leaving it unclear whether oral nitrates have the same antihypertensive effect in patients with existing health conditions. Luckily, subsequent studies have indicated that the answer is yes. For instance, in patients with chronic obstructive pulmonary disease, oral nitrates also exert a blood pressure-lowering effect [6, 7]. In a clinical study of patients with hypertension, an intervention involving oral nitrate administration (one 250 mL bottle of beetroot juice per day for 4 weeks) was implemented. Blood pressure was measured using three different methods, and oral nitrates were found to have a significant antihypertensive effect, which also improved endothelial function and reduced arterial stiffness [8]. Reduced nicotinamide adenine dinucleotide phosphate (NADPH) oxidase activity in renal vascular endothelial cells is the main target through which nitrates exert their antihypertensive effect [9].

Since NO plays a role in dilating blood vessels and inhibiting platelet aggregation and displays anti-inflammatory property, it is indispensable for maintaining vascular endothelial function. Flow-mediated dilation (FMD) is the most commonly used method for evaluating endothelial function, in which the ischemic state is revealed by measuring the diameter of the anterior wall artery. After ingesting sodium nitrate (0.15 mmol/kg), the FMD response level in healthy individuals was found to improve [10]. In addition, short-term administration of nitrates (0.45 g/d for 3 days) has been found to increase vascular dilation in the skin [11].

11.1.2 Exercise Capacity

The improved exercise capacity induced by oral nitrates is achieved by increasing vasodilation in skeletal muscles and modulating oxygen consumption (VO_2). Exercise can cause muscle hypoxia, which is conducive to the conversion of nitrites to NO. NO is recognized for its ability to dilate blood vessels and regulate mitochondrial function. There is a correlation between oral nitrate administration and oxygen consumption during exercise [12]. In a crossover-design, randomized controlled study involving healthy volunteers, loading nitrate [8.5 mg/(kg·d) for 3 days] was found to reduce oxygen consumption during exercise. This finding has been confirmed by several clinical studies in which beetroot juice was used as an intervention [13–16]. In a study involving patients with chronic obstructive pulmonary disease, those who consumed a high-nitrate beverage (140 mL of beetroot juice, containing 12.9 mmol of nitrate) exhibited better exercise capacity than those who consumed a low-nitrate beverage [6]. Supplementation with nitrates can improve the dilation of blood vessels during low-oxygen exercise in elderly individuals, but this effect was less pronounced in young individuals [17]. Additionally, oral administration of nitrate (8.2 mmol/d) not only reduced oxygen consumption during exercise but also improved muscle recovery [18]. Nitrate ingestion can further enhance muscle speed and strength [19].

In another study, healthy volunteers were recruited to consume nitrate-rich beetroot juice (70 mL), and quantitative functional magnetic resonance imaging was used to analyze oxygen utilization and skeletal muscle recovery. The results showed that nitrates could help improve energy utilization in skeletal muscle [20]. In a hypoxic state, the ingestion of nitrates could increase pulmonary oxygen uptake and enhance the endurance of muscles during high-intensity exercise [21]. Moreover, one study found that ingesting sodium nitrate could reduce the basal metabolic rate of healthy volunteers [22]. In a double-blind, randomized controlled clinical study, high-intensity exercise elevated the body's metabolic rate, and nitrate ingestion accelerated oxygen metabolism and increased exercise endurance [23].

11.2 Limitations of Nutritional Supplements with Nitrates as Their Main Component

An increasing number of clinical studies have confirmed the efficacy of nutritional supplements with nitrates as their main component. However, as a nutritional supplement, nitrates still have several shortcomings. For example, natural vegetable juices typically contain relatively low levels of nitrates, resulting in low efficiency in the elevation of nitrate levels in the body. Moreover, the nitrates in vegetable juices are rapidly metabolized in the body, making it difficult to maintain a stable nitrate concentration in the circulation and resulting in low nitrate utilization. To address these issues, a new nitrate supplementation method, "Nai Rui Te", has been developed.

Compliance with Ethical Standards This chapter is a review of previously published accounts; as such, no animal or human studies were performed.

References

1. Webb AJ, Patel N, Loukogeorgakis S, et al. Acute blood pressure lowering, vasoprotective, and antiplatelet properties of dietary nitrate via bioconversion to nitrite. Hypertension. 2008;51(3):784–90.
2. Kapil V, Milsom AB, Okorie M, et al. Inorganic nitrate supplementation lowers blood pressure in humans: role for nitrite-derived NO. Hypertension. 2010;56(2):274–81.
3. Velmurugan S, Kapil V, Ghosh SM, et al. Antiplatelet effects of dietary nitrate in healthy volunteers: involvement of cGMP and influence of sex. Free Radic Biol Med. 2013;65:1521–32.
4. Liu AH, Bondonno CP, Croft KD, et al. Effects of a nitrate-rich meal on arterial stiffness and blood pressure in healthy volunteers. Nitric Oxide. 2013;35:123–30.
5. Hobbs DA, Kaffa N, George TW, et al. Blood pressure-lowering effects of beetroot juice and novel beetroot-enriched bread products in normotensive male subjects. Br J Nutr. 2012;108(11):2066–74.
6. Kerley CP, Cahill K, Bolger K, et al. Dietary nitrate supplementation in COPD: an acute, double-blind, randomized, placebo-controlled, crossover trial. Nitric Oxide. 2015;44:105–11.
7. Berry MJ, Justus NW, Hauser JI, et al. Dietary nitrate supplementation improves exercise performance and decreases blood pressure in COPD patients. Nitric Oxide. 2015;48:22–30.
8. Kapil V, Khambata RS, Robertson A, et al. Dietary nitrate provides sustained blood pressure lowering in hypertensive patients: a randomized, phase 2, double-blind, placebo-controlled study. Hypertension. 2015;65(2):320–7.
9. Gao X, Yang T, Liu M, et al. NADPH oxidase in the renal microvasculature is a primary target for blood pressure-lowering effects by inorganic nitrate and nitrite. Hypertension. 2015;65(1):161–70.
10. Heiss C, Meyer C, Totzeck M, et al. Dietary inorganic nitrate mobilizes circulating angiogenic cells. Free Radic Biol Med. 2012;52(9):1767–72.
11. Keen JT, Levitt EL, Hodges GJ, et al. Short-term dietary nitrate supplementation augments cutaneous vasodilatation and reduces mean arterial pressure in healthy humans. Microvasc Res. 2015;98:48–53.
12. Larsen FJ, Weitzberg E, Lundberg JO, et al. Effects of dietary nitrate on oxygen cost during exercise. Acta Physiol (Oxf). 2007;191(1):59–66.
13. Bailey SJ, Fulford J, Vanhatalo A, et al. Dietary nitrate supplementation enhances muscle contractile efficiency during knee-extensor exercise in humans. J Appl Physiol (1985). 2010;109(1):135–48.
14. Bailey SJ, Winyard P, Vanhatalo A, et al. Dietary nitrate supplementation reduces the O_2 cost of low-intensity exercise and enhances tolerance to high-intensity exercise in humans. J Appl Physiol (1985). 2009;107(4):1144–55.
15. Lansley KE, Winyard PG, Fulford J, et al. Dietary nitrate supplementation reduces the O_2 cost of walking and running: a placebo-controlled study. J Appl Physiol (1985). 2011;110(3):591–600.
16. Larsen FJ, Weitzberg E, Lundberg JO, et al. Dietary nitrate reduces maximal oxygen consumption while maintaining work performance in maximal exercise. Free Radic Biol Med. 2010;48(2):342–7.
17. Casey DP, Treichler DP, Ganger CT 4th, et al. Acute dietary nitrate supplementation enhances compensatory vasodilation during hypoxic exercise in older adults. J Appl Physiol (1985). 2015;118(2):178–86.
18. Vanhatalo A, Jones AM, Blackwell JR, et al. Dietary nitrate accelerates postexercise muscle metabolic recovery and O_2 delivery in hypoxia. J Appl Physiol (1985). 2014;117(12):1460–70.
19. Coggan AR, Leibowitz JL, Kadkhodayan A, et al. Effect of acute dietary nitrate intake on maximal knee extensor speed and power in healthy men and women. Nitric Oxide. 2015;48:16–21.
20. Bentley R, Gray SR, Schwarzbauer C, et al. Dietary nitrate reduces skeletal muscle oxygenation response to physical exercise: a quantitative muscle functional MRI study. Physiol Rep. 2014;2(7):e12089.

21. Kelly J, Vanhatalo A, Bailey SJ, et al. Dietary nitrate supplementation: effects on plasma nitrite and pulmonary O_2 uptake dynamics during exercise in hypoxia and normoxia. Am J Physiol Regul Integr Comp Physiol. 2014;307(7):R920–30.
22. Larsen FJ, Schiffer TA, Ekblom B, et al. Dietary nitrate reduces resting metabolic rate: a randomized, crossover study in humans. Am J Clin Nutr. 2014;99(4):843–50.
23. Breese BC, McNarry MA, Marwood S, et al. Beetroot juice supplementation speeds O_2 uptake kinetics and improves exercise tolerance during severe-intensity exercise initiated from an elevated metabolic rate. Am J Physiol Regul Integr Comp Physiol. 2013;305(12):R1441–50.

Research and Development of New Drugs and Determination of Their Benefits

12

Yuji Wang

The components, efficacy, and limitations of nutritional supplements with nitrates as their main component were discussed in the previous section. Nitrates are hydrophilic salts derived from both endogenous and exogenous sources and are directly absorbed through the intestinal mucosa without first-pass metabolism in the liver [1]; thus, their potential for clinical applications is promising. Nitrates exist in two forms: inorganic and organic. Clinical drugs based on nitrate metabolites (including the vasodilator NO), such as nitroglycerin for the treatment of angina pectoris, have rapid effects and low biocompatibility and are associated with side effects such as headache, bradycardia, flushing, nausea, and vomiting, which limit their use. Therefore, novel nitrates may have wider clinical application prospects [2–5]. A key point is thus how to develop nitrates into a new drug.

Previous studies have shown that nitrates, as ionic compounds, play important roles in the prevention and treatment of many diseases by activating multiple signaling pathways and participating in the regulation of multiple functions. For example, they help prevent radiation-induced salivary gland damage [6, 7], ischemia-reperfusion injury [8], and irradiation injury [9, 10]. In addition, research has shown that a lack of dietary sodium nitrate in 22-month-old mice was associated with the development of metabolic syndrome and vascular endothelial dysfunction [11]. Sodium nitrate is ubiquitous in the daily diet, especially in green leafy vegetables, and it is absorbed in the gastrointestinal tract and excreted via the kidneys. Under pathological conditions, poor biocompatibility results in very limited nitrate and NO absorptions, which can also cause more severe pathological changes. In such cases, exogenous nitrate intake is necessary [12]. However, the intake of nitrate in the daily diet is insufficient to achieve therapeutic effects, and maintaining long-term effective plasma concentrations is difficult [1]. Although some nitrates can be

Y. Wang (✉)
School of Pharmaceutical Sciences, Capital Medical University, Beijing, China
e-mail: wangyuji@ccmu.edu.cn

S. Wang (ed.), *Inorganic Nitrate and Human Body Homeostasis*,
https://doi.org/10.1007/978-981-95-2824-0_12

reabsorbed [13], their concentration in the body remains relatively low, making it still difficult to reach therapeutically effective concentrations.

These studies suggested that nitrates have potential as agents for disease prevention or treatment, but their short half-life limits their use in clinical settings. Professor Wang's group is committed to improving the bioavailability of nitrates for clinical applications. One approach involves developing slow-release technology, and the other is to promote nitrate absorption (bioavailability) through the use of biocompatible complexes.

12.1 Research and Development of Nitrators

12.1.1 Artificial Intelligence (AI) Technology

Faced with the challenge of identifying traditional combination drugs through large-scale high-throughput biological experiments, a combination drug was identified using a prediction system based on AI technology. The combination drug contained vitamin C, which can form a complex with nitrates, and was identified via a three-step procedure (Fig. 12.1).

The AI model was trained directly on experimental data, yielding a series of drugs that may be compatible with nitrates, as described below.

Step 1: Establish a new drug-drug network (DDN). The creation of a DDN was the first part of the AI algorithm, designed to realize a compatible drug prediction system based on group learning. The DDN used the machine learning algorithm

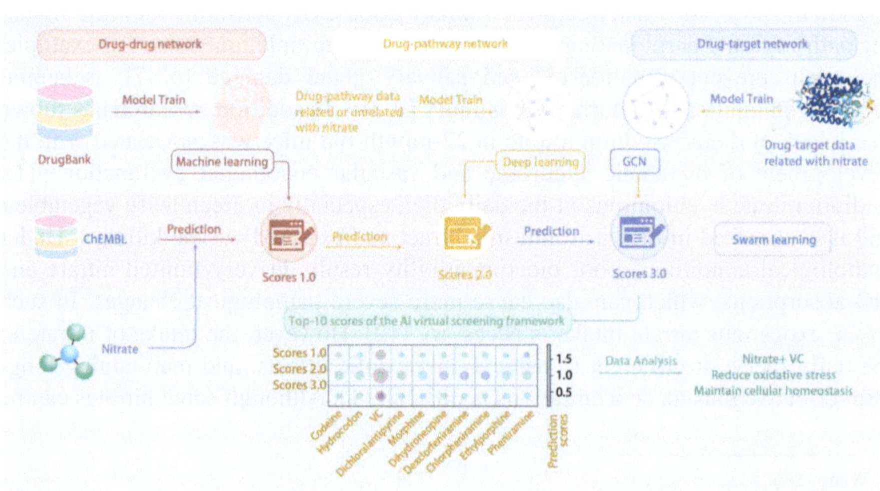

Fig. 12.1 Comprehensive schematic diagram of the swarm learning-based combination drug prediction system (SLCDPS) based on group learning SLCDPS was applied to complete the prediction scoring process for 12,000 drugs in the drug database ChEMBL

XGBoost, which utilized the drug structure data [14], to investigate drug-drug responses in humans and mice. The hyperparameters predicted by the model, such as the learning rate, were obtained by fivefold cross-validation of the practice data [15]. The performance of the drug dataset was determined on the basis of the Drug Database (DrugBank). The practice model was subsequently used to screen all the drug candidates, and good drug candidates received a score of 1.0. Using uniform manifold approximation and projection (UMAP) analysis, we found that the top 1000 DDN drug candidates were centrally distributed.

Step 2: Conduct a drug-pathway network (DPN) screening. The lack of drug pathway data related to the novel DDN made it difficult to use the network to predict which drugs would be suitable for combination with nitrates. This issue was addressed by the DPN using other data. To identify the potential mechanism of action of drug candidates, we collected nitrate information from DrugBank [6, 7]. Data on pathways related and unrelated to a drug can be obtained via an efficient in-depth learning algorithm for predicting protein structures [16]. Optimization of the drug candidates resulted in a score of 2.0; in addition, the top 100 candidates exhibited a uniform distribution according to the UMAP analysis.

Step 3: Establish a drug-target network (DTN) that combines conservative information and deep learning predictions [17]. To overcome the challenge of missing drug target data, the team collected a large amount of data on drug target links from the DrugBank, such as nitrate transporter proteins, which could prevent nitrates from being rapidly excreted. A graph convolutional network (GCN) with drug-target data [18] training was used to model the DTN, and practice and validation curves were generated to assess its performance in the DTN system.

The AI computing model combines the advantages of three networks (the DDN, DPN, and DTN) to form a unified deep learning system. By training machine learning models, potential drug-drug association networks can be built. By training a deep learning model, we can learn about drug-pathway interactions and construct a drug-pathway interaction network. By using a GCN to construct a drug-target interaction network, the AI computes the screening framework for scoring the top 10 structures.

12.1.2 Controlled Release Technology

The most commonly used nitrates are of food-grade purity and can be added to food and drinking water; however, the addition of nitrate drugs to drinking water is not approved for clinical trials. Through animal experiments, it was found that a single dose of oral nitrates was quickly absorbed in the proximal small intestine and then distributed throughout the body, but 50%–100% of the nitrates would be excreted in the urine soon after ingestion, resulting in low bioavailability. In addition, there are substantial individual differences in nitrate metabolism [1, 19]. Therefore, the low biostability and poor bioavailability of nitrates limit their use as therapeutic agents. No nitrate drug with favorable absorptive traits and protective effects has yet been reported. Owing to the short half-life of nitrates in the body, our team used modern preparation technology for screening and for optimizing the controlled-release drug preparation process, aiming to achieve a more stable blood nitrate concentration and enhanced therapeutic effects in the body. The newly developed preparations, such as one using a microcapsule technology platform and a mixture of vitamin C (VC), sodium nitrate, and chitosan to generate nitrate nanoparticles named nitrator (or nanonitrator). Nitrate delivery using these nanopreparations significantly improved the efficacy of nitrate administration in both animal injury models and animal experiments and showed no toxicity in animal experiments.

Microencapsulation is a technology for embedding and sealing solids, liquids, and gases into tiny air-dense capsules to form solid particle products [20, 21]. This prevents compounds from being affected by environmental conditions and provides controlled drug release [22]. Studies have reported that inorganic materials can be used to prepare concentrated calcium ions.

To prepare nanonitrator, sodium nitrate, sodium carboxymethyl cellulose, pectin, and chitosan were purchased, and the specific preparation steps are described below.

Step 1: Preparation of the core solution. An appropriate amount of chitosan (molecular weight 3000 daltons), an appropriate amount of VC, and 5 mL of pure water were transferred into a round-bottom flask in a 35 °C water bath to completely dissolve the material. The mixture was then cooled in a 4 °C refrigerator, and the final concentration of sodium nitrate was 4 mg/mL.

Step 2: Preparation of the microcapsule material mixture. Firstly, an appropriate amount of sodium carboxymethyl cellulose was dissolved in 5 mL of pure water, and the mixture was stirred until it completely dissolved (solution A). Then, an appropriate amount of pectin was stirred with pure water in a 50 °C water bath to obtain solution B. Finally, solutions A and B were mixed and stirred, and the total mass ratio concentration of the capsule material was approximately 3% in water.

Step 3: Preparation of the microcapsules. The capsule material and the core solution (5 mL of core material solution +9.7 mL of wall material solution) were dispersed with a high-speed homogenizer and mixed thoroughly three times. The mixture was poured into a Petri dish, frozen at −80 °C for 12 to 24 h, and then lyophilized for 12 to 24 h. The flocculant sample was milled into a fine powder by an ultrafine powder air mill, yielding the final nanonitrator for use.

12.2 Advantages of the New Nitrator

Through AI screening, we identified the best nitrate-VC combination (new nitrator) from the database of 12,000 drugs. To overcome the disadvantages of low bioavailability and short half-life, we developed a nanonitrator containing a sodium nitrate-VC nanoformulation that can be taken orally. After synthesis, optimization, and characterization of the agent's physicochemical properties, we found that the nanoparticles could provide controllable release of nitrates *in vitro* and *in vivo* while improving absorption and bioavailability.

In addition, because it could significantly upregulate PI3K-AKT signaling, the new nitrator was shown to have a superior effect in maintaining cellular homeostasis compared to nitrates, indicating its potential for clinical application. This nanotechnology provides a method for loading inorganic compounds into slow-release nanoparticles, providing new possibilities for using inorganic molecules in clinical applications, and preclinical drug studies are underway.

A series of experimental studies have demonstrated that slow-release formulations enhance the body's absorption of nitrates and improve bioavailability. In animal models of salivary gland radiation damage and liver ischemia-reperfusion, these formulations have shown promise in preventing and treating both oral and systemic diseases, indicating their potential for clinical application.

To evaluate the safety of nanonitrator, additional studies on the preparation process and safety assessments in accredited laboratories are needed. Morphological and functional evaluations revealed that the nanonitrator exerted certain effects in animal models of salivary gland radiation damage and liver ischemia-reperfusion. Organ and blood samples were collected for histological examination and biochemical index testing. The specific evaluation experiments were conducted as follows.

A capillary straw was used to collect saliva from rats. One end of the capillary straw was positioned at the bottom of the inlet, while the other end was placed at the bottom of a 1.5 mL centrifuge tube. The upper and lower mandibular incisors were used to secure the tube against the wall, and saliva was collected for 20 min. The amount of saliva was then quantified using the weighing method and recorded.

To establish a salivary gland radiation injury model, a group of 5 rats were anesthetized and placed in the supine position on an accelerator treatment bed. Each rat's submandibular gland area was covered with a 3.0 cm × 3.0 cm, 1 cm-thick wax film and a water-equivalent material. The total length of the submandibular gland was approximately 3.0 cm. A 2D source-to-surface distance irradiation technique was employed. The average radiation energy was 6 MV, with a dose rate of 300 cGy/min. The irradiation field measured 34 cm × 3.0 cm, and the radioactive source was positioned 100 cm above the upper surface of the wax film. Using the EQD2 equation for conversion, a single irradiation dose of 15 Gy was equivalent to a fractionated dose of 2 Gy per fraction, with a total of 16 fractions. This resulted in a total biological effect of 31.25 Gy. The beams were centered on the submandibular glands of the five rats and were delivered at a gantry angle of 0°.

After the experimental animals were euthanized, they were placed in the supine position and disinfected, and a neck incision was made to completely remove the submandibular glands. The glands were washed, cut into 0.5 cm × 0.5 cm pieces, and fixed in 4% paraformaldehyde (pH 7.2) at 4 °C for 24 to 48 h. The tissue slices were then embedded using water immersion wax, sectioned to a thickness of 4 μm, heated (dewaxed), and stained with hematoxylin-eosin (HE) for microscopic examination. To evaluate nitrate absorption, the levels of NO in the salivary gland tissues were also measured. Compared with the other groups, the radiotherapy group presented a decreasing trend, with a significantly greater incidence of resistance (Fig. 12.2a). Histological examination of paraffin-embedded tissue blocks at 8 weeks post-radiotherapy revealed the presence of a substantial number of vacuoles within the cytoplasm of the submandibular gland acinar cells in the radiotherapy group. Compared with that in the radiotherapy group, the acinar cell vacuolar area was diminished in all the nitrate treatment groups. Notably, the number of vacuoles was significantly lower in the nitrator group than in the nitrate group (Fig. 12.2b). Since the primary function of the salivary gland is to secrete saliva, a study involving a single salivary gland irradiation dose of 15 Gy revealed that, from one week prior to radiation until eight weeks post-radiation, the saliva flow rate in the rats in the radiation group decreased to approximately 50% of pre-radiation levels. Both the nitrate group and the radiation group showed varying degrees of improvement, with the nanonitrator group demonstrating the most significant improvement, followed by the nitrate group (Fig. 12.2c).

While the animals were under anesthesia, ischemia was induced, and the incision was sutured after one hour. The experimental animals were treated six hours later, and tissue samples were collected for relevant testing. For the hepatic ischemia-reperfusion plus nitrate group, normal mice received 0.75 mmol/kg sodium nitrate for five days, followed by a 12-h fasting period. Anesthesia was then induced, and open surgery along the abdominal midline was performed to induce ischemia. After one hour, the animals were sutured, and six hours later, the tissue samples were collected for relevant testing.

The modeling of liver ischemia and reperfusion injury involved peritoneal injection of heparin into anesthetized mice to prevent blood coagulation. A laparotomy along the abdominal midline was conducted to fully expose the liver. The hepatic artery and portal vein were occluded with a nontraumatic vascular clip to induce partial hepatic ischemia in the left and middle lobes of the liver. Throughout the procedure, the mice were kept on a heating pad to maintain a body temperature of 37 °C. After one hour, the vascular clip was removed to allow reperfusion. Six hours post-suturing, plasma and liver tissues were collected for testing.

After deep anesthesia, the mice were euthanized, and a midline laparotomy was conducted. Saline was perfused into the liver via the aorta, hepatic artery, or portal vein. The liver was then isolated and placed in normal saline for subsequent analysis.

To investigate the effects of nitrates and the nitrator on liver ischemia-reperfusion injury, the levels of NO in both plasma and liver tissue were measured, which were significantly higher in the mice pretreated with the nanonitrator and nitrate compared to those in the ischemia-reperfusion group. Additionally, NO levels in the group pretreated with the nanonitrator were also significantly higher than those in the nitrate-alone group (Fig. 12.3a).

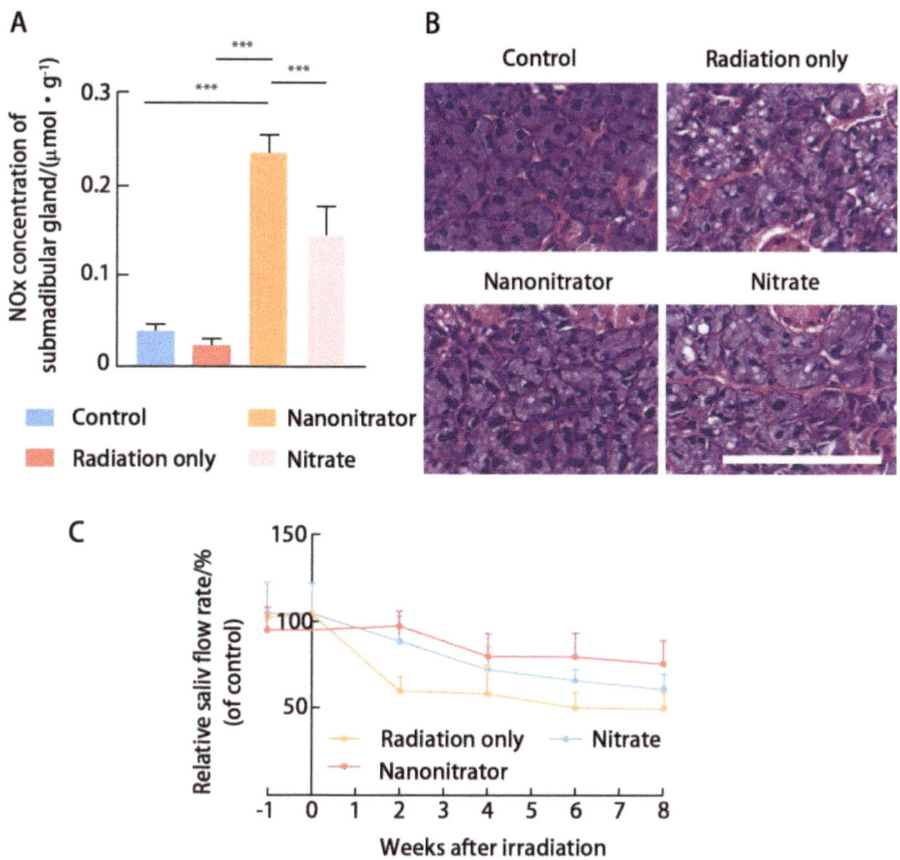

Fig. 12.2 Evaluation of radiation damage in the salivary glands in the rat model. (**a**) NO in the submandibular glands at 8 weeks after radiotherapy. (**b**) H&E staining of the submandibular glands; scale bar = 100 μm. (**c**) Irritant-induced saliva production at different time points before and after radiotherapy

Pathology analysis revealed that liver ischemia-reperfusion injury induced marked tubular necrosis and inflammatory cell infiltration, accompanied by morphological changes in the liver tissue. Nitrator and nitrate pretreatment reduced these pathological changes in the liver caused by ischemia-reperfusion injury; however, the nitrator more significantly improved the degree of ischemia-reperfusion caused by liver damage (Fig. 12.3b).

Plasma ALT and AST, biomarkers of hepatic function, were significantly elevated in the ischemia-reperfusion group compared to the control group. M5 mice pretreated with the nanonitrator and exhibited significantly lower levels of both ALT and AST compared to mice in the ischemia-reperfusion only group (Fig. 12.3c).

To further investigate the clinical translation potential, we simultaneously conducted an early mechanistic study to evaluate the cell states and mRNA transcript

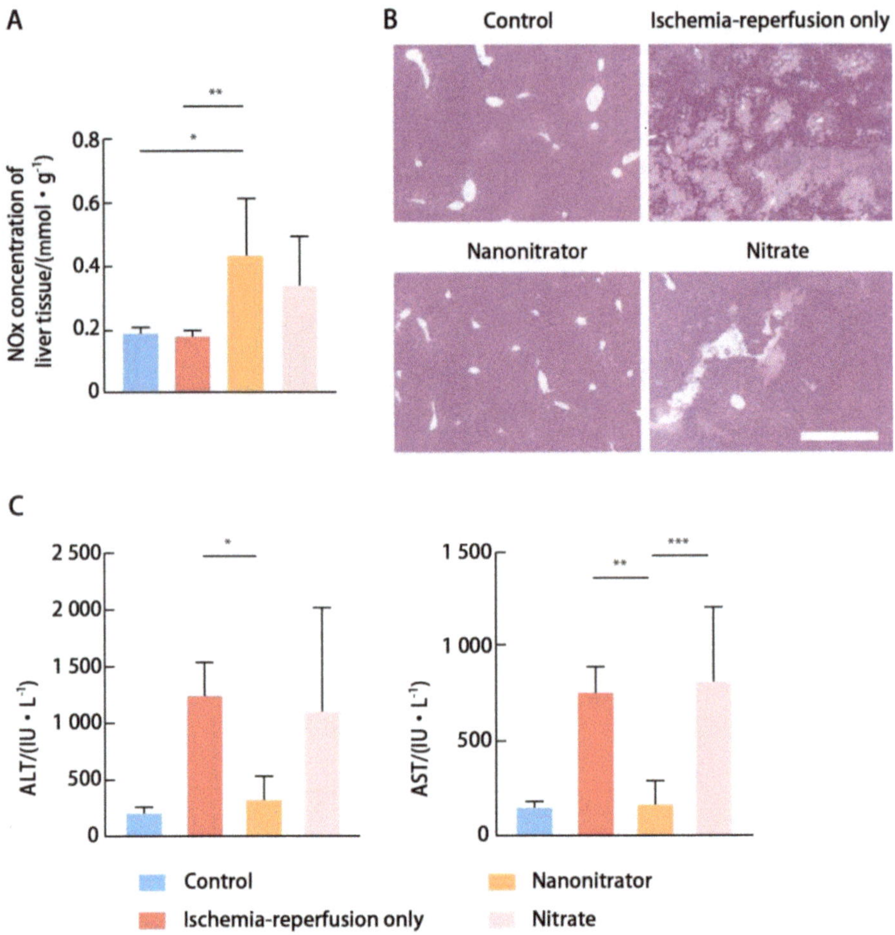

Fig. 12.3 Evaluation of the mouse liver ischemia-reperfusion model. (**a**) NO concentration in the liver tissue. (**b**) H&E staining of liver tissue; scale bar = 200 μm. (**c**) Plasma alanine aminotransferase (ALT) levels and plasma aspartate aminotransferase (AST) levels

levels during radiotherapy. The experimental results from the pharmacodynamic evaluations of the two animal models suggested that the nanonitrator had significant effects in animal models of salivary gland radiation damage and hepatic ischemia-reperfusion.

12.3 Safety Evaluation

Drug safety evaluation is a problem-based assessment (question-based review), which involves comprehensive investigations designed for clinical research objectives. Such evaluation is performed on the basis of a complete understanding of the subject matter and in strict compliance with China's measures for the administration of drug registration, relevant technical guidelines for safety evaluation, and applicable norms, including good laboratory practice (GLP), laboratory animal regulations, and biosafety protocols. On the basis of the project's approval and the objectives of innovative drug research and development, and in accordance with the principles of specific problem-by-case analysis, a preclinical safety evaluation is designed to obtain meaningful safety evaluation data to explore or evaluate the approval or rejection of innovative drugs from the perspective of safety and to reduce the risk of new drugs and clinical research.

For chemical drugs, appropriate test methods should be selected according to the structural characteristics and physical and chemical properties of the test articles, data from similar compounds or marketed drugs, pharmacological efficacy profiles, indications, target population characteristics, and testing objectives with respect to the test articles. An appropriate comprehensive approach should be considered in designing the testing scheme.

For safety evaluation in preclinical studies, particularly those involving resistance, the following factors should be taken into account: application project, test candidates, indications, mode of administration, animal models, analytes, and test methods.

Based on the structural characteristics, physicochemical properties, and pharmacological efficacy profiles of the test articles, the following experimental items for safety evaluation have been designed:

1. *In vitro* pharmacokinetic testing
2. *In vivo* pharmacokinetic testing

    2.1. Single-dose pharmacokinetic testing in rats
    2.2. Repeat-dose pharmacokinetic testing in rats
    2.3. Urine and fecal excretion tests in rats
    2.4. Rat bile excretion test
    2.5. Tissue distribution test in rats
    2.6. Single-dose pharmacokinetic testing in beagle dogs
    2.7. Seven-day repeat administration pharmacokinetic testing in beagle dogs

3. Safety evaluation

    13.1 Determination of maximum tolerated dose and 14-day dose range in rats
    13.2 Determination of maximum tolerated dose and 14-day dose range in beagle dogs
    13.3 Four-week administration plus four-week recovery toxicology testing in rats
    13.4 Four-week administration in beagle dogs plus four-week recovery toxicology testing
    13.5 Functional observation battery (FOB) test in rats
    13.6 Respiratory tests in rats
    13.7 Cardiovascular system testing in beagle dogs
    13.8 *hERG* (human ether-à-go-go-related gene) testing

   13.9 Bacterial reverse mutation assay
   13.10 Chromosomal aberration test in Chinese hamster ovary cells
   13.11 *In vivo* micronucleus test
4. Program and report preparation
5. Electronic data submission

Note: The design is intended to evaluate the nanomedicine-based formulation in the drug, and additional evaluation and adjustment of analysis methods will be required based on the nano characteristics.

Example 12.1: A Seven-Day Repeated-Dose Pharmacokinetic Testing in Beagle Dogs

Principles to follow		Guidelines of China NMPA and the US FDA
GLP		Insufficient GLP
Animal information	Species	Beagle dogs (administration history)
	Test cycle	8 days
	The number of animals	6; half male and half female
	Group count	1
	Total number of animals	6
Distribution and analysis	Dosage	1 time
	Frequency of administration	Once daily
	Route of administration	Presumably oral
	Sample analysis of dispensing drugs	Indeterminate
Observations	Cage observations	2 times a day
	Detailed clinical observations	1 time before animal administration
	Weight	1 time before animal administration
	Amount of food	Indeterminate
Sample collection	Pharmacokinetic samples collection	Eight spots were collected each within 24 h on day 1 and 7, and 24 h after dosing on day 4 and 5
	Pharmacokinetic sample analysis	There were 108 samples, assuming 1 sample for LC-MS/MS method
Anatomy and pathology	Anatomy/tissue preservation	No dissection
Report and archive	Report	Non-GLP Chinese report
	Keep in the archives	Five years

Example 12.2: An In Vitro Chromosome Aberration Test in Chinese Hamster Ovary Cells

Principles to follow		Guidelines of China NMPA and the US FDA
GLP		GLP
Test system information	Experimental system	Chinese hamster ovary cells
	Test cycle	4 weeks
	Metabolic activation system	The rat liver was homogenized at a 1.5% concentration in medium
	Solvent/positive control material	In the main experiment, parallel solvent/positive controls were set up under metabolically activated or non-metabolically activated conditions
	Methods	Then, cells were incubated with the articles under two conditions: for 3 h or for 3 h and 20 h. The cells were harvested after 20–24 h. Two hours before cell harvesting, Colcemid was added to the medium to accumulate metaphase phase cells
Analysis	Preparation	For ready to use, at least 2 times
	Frequency of administration	1 time
	Sample analysis of preparations	1 time in the main experiment
Preliminary experiment	Number of dose groups	At least 8 doses were used.
	Number of cell vials	One vial of cells per dose
	Evaluation metrics	Cytotoxicity and the solubility of the articles in the culture medium
The main experiment	Experimental contents	Based on the preliminary experimental outcomes, a minimum of three doses were chosen, with the uppermost concentration primarily dictated by the test substance's toxicity and/or solubility in the cells. For soluble, nontoxic test substances, the maximum test concentration typically did not exceed 0.5 mg/mL or 1 mM, whichever was lower. In the case of soluble, cytotoxic substances, the highest concentration was established based on the extent of cytotoxicity, with a general guideline of 55% ± 5% cell viability. For insoluble substances, the highest concentration was usually set at the minimum precipitation level. Chromosomal structural aberrations and numerical aberrations, including polyploidy and endoreplication, were documented separately. All slides were coded in a blinded manner prior to the analysis of chromosomal aberrations. A minimum of 300 metaphase cells were counted at each dose to evaluate the frequency of aberrant cells

Phase 1: Preclinical research phase. Basic research and drug discovery were completed, drug targets were identified, new compounds with therapeutic potential were discovered. The pharmacological effects of the drugs are being studied, and initial development of the preparations is underway. Prior to conducting clinical trials, the preclinical study data for the test drug, including prescription composition, manufacturing process, and quality test results, must be submitted to and approved by the China National Medical Products Administration.

Phase 2: Drug clinical trials. Upon approval of the clinical trial, the second phase commences. The scope of clinical trials, as stipulated in the measures for drug registration, encompasses research on humans aimed at determining the safety and efficacy of a drug. Clinical trials are categorized into phase I through phase IV and bioequivalence trials, which include clinical pharmacology studies, exploratory clinical trials, confirmatory clinical trials, and postmarketing studies. Phase I serves as a safety assessment, typically involving 20 to 100 healthy volunteers to collect safety data and determine a safe dosage range for the drug. Concurrently, efficacy and safety are evaluated by evaluating the absorption, distribution, metabolism, and elimination of the drug within the human body, as well as tolerance levels, thereby establishing a foundation for the drug's administration strategy. The rigor of this testing phase is directly correlated with the success or failure of drug development. The clinical trial period for drug efficacy typically involves 100 to 500 patients with related conditions, primarily to collect efficacy data for drug treatment. This phase can be expanded to include efficacy tests, often necessitating a larger cohort of clinical and hospitalized patients. These tests can be conducted across multiple medical centers under strict medical supervision to further ascertain the drug's effectiveness and side effects, as well as its interactions with other medications. Consequently, the clinical trial period is the most critical phase in the entire process.

Stage 3: Postmarketing surveillance. Following the marketing phase, the clinical trial enters the postmarketing surveillance stage, which focuses on observing the drug's efficacy and adverse reactions in broader populations. Given the limited effectiveness of conventional drug therapies and radiotherapy, the prognosis remains grim, highlighting the pressing need for novel treatment strategies or combination therapies. Currently, preclinical research on nanonitrator indicates promising efficacy and safety profiles, demonstrating substantial therapeutic effects in two models. Notably, there is a lack of similar research in China. The findings from Wang's research group suggested that nanonitrator could be developed into a new adjunctive therapeutic drug for clinical use, potentially playing a significant role in the prevention and treatment of related diseases.

Compliance with Ethical Standards All applicable international, national, and/or institutional guidelines for the care and use of animals were followed.

References

1. Omar SA, Artime E, Webb AJ. A comparison of organic and inorganic nitrates/nitrites. Nitric Oxide. 2012;26(4):229–40.
2. Jones AM, Thompson C, Wylie LJ, et al. Dietary nitrate and physical performance. Annu Rev Nutr. 2018;38:303–28.
3. Bondonno CP, Blekkenhorst LC, Liu AH, et al. Vegetable-derived bioactive nitrate and cardiovascular health. Mol Aspects Med. 2018;61:83–91.
4. Parker JO, Nitrate tolerance. A problem during continuous nitrate administration. Eur J Clin Pharmacol. 1990;38(Suppl 1):S21–5.
5. Bode-Böger SM, Kojda G. Organic nitrates in cardiovascular disease. Cell Mol Biol (Noisy-le-grand). 2005;51(3):307–20.
6. Feng X, Wu Z, Xu J, et al. Dietary nitrate supplementation prevents radiotherapy-induced xerostomia. Elife. 2021;10:e70710.
7. Li S, An W, Wang B, Li J, et al. Inorganic nitrate alleviates irradiation-induced salivary gland damage by inhibiting pyroptosis. Free Radic Biol Med. 2021;175:130–40.
8. Li S, Jin H, Sun G, et al. Dietary inorganic nitrate protects hepatic ischemia-reperfusion injury through NRF_2-mediated antioxidative stress. Front Pharmacol. 2021;12:634115.
9. Chang S, Hu L, Xu Y, et al. Inorganic nitrate alleviates total body irradiation-induced systemic damage by decreasing reactive oxygen species levels. Int J Radiat Oncol Biol Phys. 2019;103(4):945–57.
10. Wang W, Hu L, Chang S, et al. Total body irradiation-induced colon damage is prevented by nitrate-mediated suppression of oxidative stress and homeostasis of the gut microbiome. Nitric Oxide. 2020;102:1–11.
11. Kina-Tanada M, Sakanashi M, Tanimoto A, et al. Long-term dietary nitrite and nitrate deficiency causes the metabolic syndrome, endothelial dysfunction and cardiovascular death in mice. Diabetologia. 2017;60(6):1138–51.
12. Park JW, Piknova B, Huang PL, et al. Effect of blood nitrite and nitrate levels on murine platelet function. PLoS One. 2013;8(2):e55699.
13. Weitzberg E, Lundberg JO. Novel aspects of dietary nitrate and human health. Annu Rev Nutr. 2013;33:129–59.
14. Bryant DH, Bashir A, Sinai S, et al. Deep diversification of an AAV capsid protein by machine learning. Nat Biotechnol. 2021;39(6):691–6.
15. Identification of antimicrobial peptides from the human gut microbiome using deep learning. Nat Biotechnol. 2022;40(6):838–9.
16. Boström J. Transformers for future medicinal chemists. Nat Mach Intell. 2021;3:102–3.
17. Vaishnav ED, de Boer CG, Molinet J, et al. The evolution, evolvability and engineering of gene regulatory DNA. Nature. 2022;603(7901):455–63.
18. Schulte-Sasse R, Budach S, Hnisz D, et al. Integration of multiomics data with graph convolutional networks to identify new cancer genes and their associated molecular mechanisms. Nat Mach Intell. 2021;3:513–26.
19. EFSA Panel on Food Additives and Nutrient Sources added to Food (ANS); Mortensen A, Aguilar F, et al. Re-evaluation of sodium nitrate (E 251) and potassium nitrate (E 252) as food additives. EFSA J, 2017, 15(6):e04787.
20. Wang X, Zhang M, Xing F, et al. Effect of a healing agent on the curing reaction kinetics and its mechanism in a self-healing system. Appl Sci. 2018;8(11):2241.
21. Zhang X, Xu D, Jin X, et al. Nanocapsules of therapeutic proteins with enhanced stability and long blood circulation for hyperuricemia management. J Control Release. 2017;255:54–61.
22. Surnar B, Kamran MZ, Shah AS, et al. Orally administrable therapeutic synthetic nanoparticle for Zika virus. ACS Nano. 2019;13(10):11034–48.

Part V

Inorganic Nitrate and Homeostatic Medicine

Lizheng Qin

The beneficial effects of inorganic nitrates on multiple systems and organs have been widely demonstrated. Available evidence suggests that inorganic nitrates are bioactive substances that are widely found in nature and play vital roles in biological processes in living organisms. The biological processes in living organisms follow the second law of thermodynamics, exhibiting a spontaneous tendency toward disorder and increased entropy. To maintain order and a low entropy level in the system, living organisms continuously transform materials and energy in the environment from a high-entropy state to a low-entropy state for utilization while eliminating metabolic byproducts from the body. During this process, a relatively stable internal environment must be maintained, and this is known as homeostasis. Homeostasis is a self-regulated mechanism for achieving dynamic equilibrium that helps organisms maintain the relative stability of their internal environment in the ever-changing external environment, thereby supporting normal biological activities and sustaining physiological functions across various systems. Furthermore, organisms can adapt to external environmental changes and counteract physiological challenges through homeostasis regulation. Thus, homeostasis is crucial for maintaining internal stability and is beneficial for survival.

Homeostatic imbalance plays an important role in the onset and progression of numerous diseases. Consequently, restoring homeostasis is a fundamental strategy for reversing these pathological states. Homeostatic medicine integrates the regulation of homeostasis at the molecular, cellular, organ, and system levels. After studying the causes and manifestations of homeostatic imbalance during the disease process, homeostatic medicine can be applied to improve organ function and prevent or treat disease by targeting key aspects of homeostasis regulation. The nitrate-nitrite-nitric oxide (NO) pathway is important for the maintenance of homeostasis and complements the endogenous NO pathway. In addition, the interaction between

L. Qin
L.Q., Beijing Stomatological Hospital, Capital Medical University, Beijing, China

sialin and inorganic nitrates contributes to systemic homeostasis by participating in the regulation of various aspects of cellular homeostasis. Therefore, the nitrate-sialin feedback loop and its role in homeostasis represent promising research topics for future studies on homeostasis regulation.

How Inorganic Nitrate Maintains Homeostasis

13

LiZheng Qin

13.1 The Nitrate-Sialin System and the Maintenance of Homeostasis

NO is crucial for the homeostasis regulation of multiple systems in the body. Inorganic nitrates can regulate NO *in vivo* through the nitrate-nitrite-NO pathway, thereby playing an important physiological role in the maintenance of homeostasis. Sialin serves as a nitrate transporter, transporting inorganic nitrates into cells and organelles to facilitate different biological functions in cells. Thus, the inorganic nitrate-sialin system represents a fundamental mechanism underlying homeostasis regulation.

13.1.1 Major Molecule Involved in Homeostasis Regulation—NO

NO is a major molecule that has a key function in maintaining homeostasis. As a signaling molecule, it helps achieve dynamic homeostasis of the neurohumoral system by modulating both the neuroendocrine and autonomic nervous systems. NO can play a protective role in conditions of homeostatic imbalance (e.g., dehydration, hemorrhage, stress) by re-establishing an appropriate autonomic nervous system-fluid balance [1]. However, it can play dual roles in certain contexts. For example, while NO can mitigate tissue inflammation by suppressing leukocyte recruitment, as inflammation progresses, excessive NO production induces oxidative stress in cells, thereby exacerbating inflammation. Thus, maintaining NO homeostasis contributes to the regulation of systemic homeostasis. NO can be synthesized endogenously via NO synthase (NOS). However, when endogenous NO synthesis is

L. Qin (✉)
Beijing Stomatological Hospital, Capital Medical University, Beijing, China
e-mail: qinlizheng@ccmu.edu.cn

S. Wang (ed.), *Inorganic Nitrate and Human Body Homeostasis*,
https://doi.org/10.1007/978-981-95-2824-0_13

compromised, such as during endothelial dysfunction, NO homeostasis can be maintained via exogenous NO pathways, particularly via the ingestion of inorganic nitrates (Fig. 13.1) [2].

13.1.2 Inorganic Nitrates Regulate Homeostasis Through the Nitrate-Nitrite-NO Pathway

Upon ingestion, nitrates are rapidly absorbed into the bloodstream in the upper gastrointestinal tract. Approximately 25% of these nitrates are taken up by salivary glands, concentrated in saliva, and secreted into the oral cavity, which represents the process known as entero-salivary circulation [3]. Nitrates in saliva are then converted to nitrites by nitrate-reducing bacteria in the oral cavity. After that, nitrites are swallowed and reenter the gastrointestinal tract, where they are absorbed into the body via the bloodstream and reduced to NO by various enzymes in blood and tissues. This sequence of events is known as the nitrate-nitrite-NO pathway [4]. These processes play crucial roles in maintaining NO homeostasis and further exert beneficial physiological effects by enhancing systemic homeostasis. For example, oral inorganic nitrates can protect the gastrointestinal tract by increasing gastrointestinal blood flow and regulating the gut microbiome, prevent obesity by influencing fat metabolism, and increase sensitivity to cisplatin-based chemotherapy to increase the effect of cancer treatment. In addition, nitrates can strongly activate various cellular signaling pathways, including the MAPK, PI3K-Akt, mTOR, and Wnt signaling pathways, and can regulate glutathione metabolism and cell cycle, thus playing important roles in cellular regeneration, metabolism, and disease progression [5].

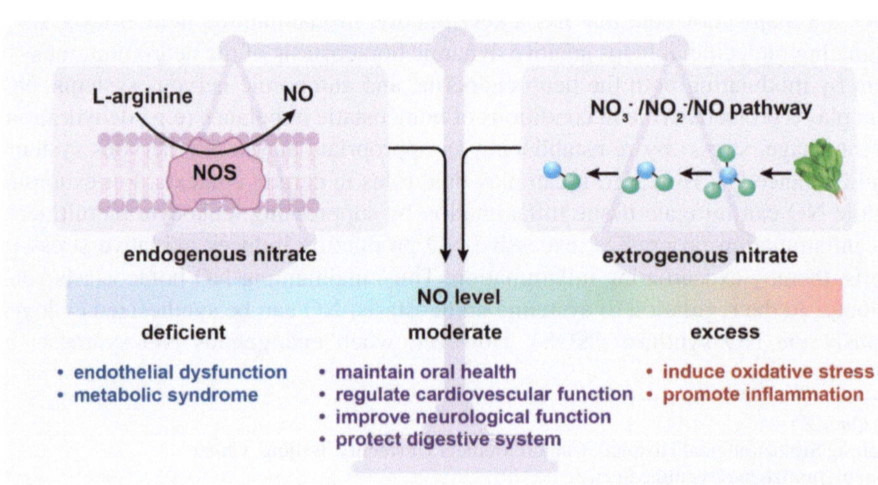

Fig. 13.1 Relationship among inorganic nitrate, NO homeostasis, and systemic homeostasis

13.1.3 Interactions Among Inorganic Nitrates, Sialin, and NO During the Maintenance of Homeostasis

Sialin functions as a nitrate transporter in mammalian cell membranes and is a key protein involved in the physiological function of nitrates and the maintenance of NO homeostasis. Notably, it is highly expressed in human salivary glands and transports nitrates into acinar cells. Therefore, the functions of salivary glands can strongly influence the physiological function of inorganic nitrates [6]. In turn, nitrates also exert modulatory effects on the functions of salivary glands. Elevated blood nitrate concentrations are usually accompanied by increased sialin expression across major organs, subsequently triggering a range of cellular biological functions. Inorganic nitrates can upregulate sialin expression in mesenchymal stem cells (MSCs) to improve their mitochondrial function and decrease their level of senescence. This bidirectional regulatory mechanism between sialin and nitrates is referred to as the nitrate-sialin feedback loop [7].

Oral nitrates prevent nonalcoholic fatty liver disease by modulating sialin expression levels in macrophages, thereby regulating the M1/M2 macrophage ratio. In rats with ovariectomy-induced xerostomia, inorganic nitrates increase the expression of the aquaporin protein AQP5 in the salivary glands and increase saliva secretion, effectively reducing fibrosis and cellular atrophy in salivary gland tissues. In a miniature pig model of radiation-induced salivary gland dysfunction, nitrate administration increased sialin expression in acinar cells, thereby facilitating the entry of nitrates into the cells. This positive feedback mechanism activates the EGFR-AKT-MAPK signaling pathway, thereby promoting the proliferation and reducing the apoptosis of acinar and ductal cells. In salivary gland cells, the nitrate-sialin feedback loop preserves cell survival following radiation injury by regulating autophagy. In the context of mitochondrial destabilization, nitrates are transported into the mitochondria by sialin in the mitochondrial membrane, which in turn reduces mitochondrial damage and improves mitochondrial function. In both alcoholic and nonalcoholic fatty liver disease, nitrates play a crucial role in modulating the balance of anti-inflammatory and proinflammatory macrophages, thereby ameliorating fatty liver. This evidence highlights the interaction between sialin and nitrates, which could be beneficial for the nitrate-mediated regulation of NO homeostasis, thereby improving systemic homeostasis. Therefore, the nitrate-sialin feedback loop and the role of nitrates in the regulation of NO homeostasis are important potential areas for future research on homeostasis regulation (Fig. 13.2).

13.2 Homeostasis and Homeostatic Medicine

Homeostasis is essential for normal biological activities in living organisms, as it facilitates the maintenance of stability in the internal environment and physiological functions in organisms. Furthermore, homeostasis constitutes a self-regulating dynamic equilibrium in living organisms, allowing them to respond to various physiological stimuli and pathogenic factors. When homeostasis is disrupted and the

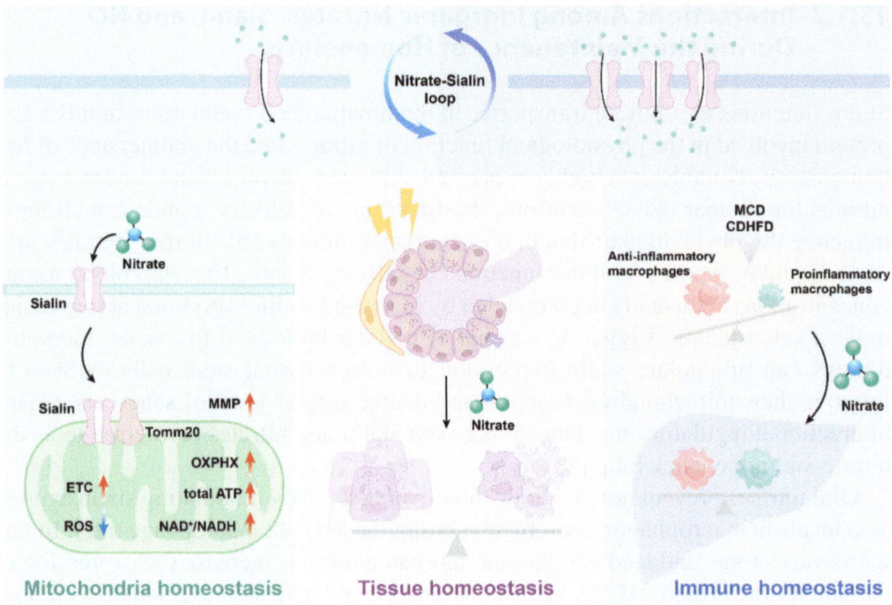

Fig. 13.2 The nitrate-sialin feedback loop and its role in regulating homeostasis

regulatory system is impaired, organisms will undergo a range of functional, structural, and metabolic alterations, which manifest as clinical signs and symptoms of disease, as well as behavioral abnormalities.

Homeostatic medicine focuses on exploring the role of homeostasis in health and disease and investigating the patterns of homeostatic equilibrium and regulatory strategies at molecular, cellular, organ, and organism levels, aiming at maintaining a healthy status and contributing to the diagnosis and treatment of relevant diseases. In contrast to conventional "symptomatic" treatments, homeostatic medicine represents a new paradigm that pays more attention to the root causes of diseases, the "etiological treatment" of homeostatic disruption, and the dynamic re-establishment of homeostasis in various systems. It offers a novel perspective for understanding health, disease diagnosis, and treatment.

13.2.1 History of Homeostasis and Homeostasis in Traditional Chinese Medicine

13.2.1.1 History of the Development of Homeostasis Theory

The development of homeostasis theory emerged from a dialectical process. The original theory of body equilibrium can be traced back to 460 BCE when Hippocrates first proposed the theory of humorism, which describes a balanced state in the human body among four humors, namely, yellow bile, blood, phlegm, and black bile. These humors are constantly produced and consumed *in vivo* and play crucial roles in maintaining health and mediating disease. Although the theory of humorism has many limitations when analyzed from a modern medical perspective, it has historically served as both an inspiration and a foundation for modern Western medicine. Hippocrates further suggested that the body can heal itself, while the role of the physician is to clear the obstacles to healing and allow the body to return to its natural state. This idea is still widely accepted today, and for this reason, Hippocrates is known as the "father of Western medicine."

As research into body functions has advanced, the concept of homeostasis regulation has become more comprehensively understood. The French physiologist Claude Bernard (1813–1878) proposed that living systems have internal stability that buffers and protects the body from the influence of the changing external environment, which he called the "stability of the milieu intérieur (internal environment)." Bernard conceptualized the human body as an assemblage of body fluids and cells in an internal environment, suggesting that the stability and relative autonomy of this internal environment are essential for the survival of the body. However, Bernard's theory was more radical, as he believed that the internal environment of the body was static and independent of changes in the external environment [1].

Bernard's theory of the "stability of the milieu intérieur" was further developed by the American physiologist Walter Bradford Cannon (1871–1945), who introduced the concept of dynamic balance. This concept describes the self-regulatory process by which organisms can maintain stability while adapting to the changing environment. In Cannon's book *The Wisdom of the Body*, he explained that homeostasis requires two components at the same time, namely, internal stability within a range and the ability to remain stable by regulating variables. Cannon's homeostasis theory has been widely recognized in modern medicine and serves as a foundational theoretical framework for ongoing research on homeostasis [2].

13.2.1.2 Homeostasis Regulation in Traditional Chinese Medicine

Similar to the concept of homeostasis in Western medicine, traditional Chinese medicine has long embraced the theory of "balance of yin and yang." *Huangdi Neijing* initially proposed the concepts of "yin and yang in relative equilibrium" and "harmonious coordination." "Yin and yang in relative equilibrium" refers to the opposing and restricting relationship between yin and yang, which eventually harmonizes to form a dynamic balance. "Harmonious coordination" emphasizes the importance of maintaining the balance between yin and yang. It is considered more important to restore balance than to eliminate the cause of an illness, and excessive treatment should be avoided since it will disrupt this balance. Thus, despite the substantial differences between Eastern and Western medicine, their ideas converge on the role of homeostasis regulation in health and disease [3].

13.2.2 Mechanisms of Homeostasis Regulation

13.2.2.1 Feedback Regulation

Feedback regulation represents a fundamental mechanism of homeostasis regulation and usually consists of four main components: ① the variable to be controlled; ② a sensor that monitors the variable; ③ a comparator or central processing unit where the information provided by the sensor is fed back into the system. The information is compared with the specified value, and ④ effectors are used to regulate the desired control variables. These components collectively form a closed feedback signaling mechanism. In a negative feedback process, the activity of the effector is opposite to the changes in the variables to buffer them. In a positive feedback process, the activity of the effector is the same as the changes in the variables to achieve rapid changes in the state of the body by amplifying a control signal. Notably, these four components are only the basic requirements for the feedback system, and the feedback system *in vivo* is far more complicated, involving the superposition and nesting of multiple feedback pathways [4].

For example, blood pressure is typically maintained within a relatively stable range through feedback regulation. The receptors involved in this homeostatic system are baroreceptors located in the aortic arch and carotid sinuses that respond to changes in arterial pressure. The solitary nucleus in the medulla oblongata of the brain processes signals that are fed back from pressure receptors and subsequently acts on the effectors located in the blood vessels and heart by modulating the neural activity of the sympathetic and parasympathetic nerves. When the blood pressure is elevated, baroreceptors are activated, and the regulation of the solitary nucleus decreases sympathetic activity and increases the vascular diameter. At the same time, increased parasympathetic activity lowers the heart rate and stroke output, resulting in a decrease in blood pressure. The opposite occurs when the blood pressure falls below an equilibrium value. This negative feedback regulation effectively buffers fluctuations in blood pressure so that the body's blood pressure can remain relatively stable all day, despite changes in the environment or behavior.

13.2.2.2 Feedforward Regulation

Feedforward regulation is another major mechanism of homeostasis regulation that refers to the assessment and adjustment of impending changes before they actually occur. Several levels of regulation are involved in homeostasis. The first level is the effector responsible for receiving higher-level regulatory signals and variables. The second level is feedback regulation, also known as autonomic regulation, which processes signals detected by receptors and initiates adjustments to the first level. The third level is located in the central nervous system, which is responsible for processing information transmitted from the second level and integrating the changing information of the environment to coordinate the physiological behavior of multiple feedback systems. This regulation can be unconscious or regulated by subjective consciousness under certain conditions [5]. Again, taking blood pressure regulation as an example, when faced with danger or a challenge, the cardiac output rate and blood pressure can be raised through the central nervous system in response to impending environmental changes. This regulation is unconscious, in contrast to the response to a decrease in the ambient temperature, wherein a person consciously puts on more clothes to regulate body temperature.

13.2.3 Concept of Homeostatic Medicine

13.2.3.1 Definition of Homeostatic Medicine

Homeostatic medicine is an academic discipline focusing on the study of homeostasis of molecules, cells, tissues, organs, and the whole system within the human body. It is a comprehensive discipline with the basic idea of maintaining human health and preventing and diagnosing diseases by maintaining homeostasis. Through systematic investigations into the patterns and mechanisms of homeostasis regulation at multiple levels, including interactions with molecules, cells, tissues, organs, the whole body, and the external environment, homeostatic medicine proposes various methods and strategies to guide clinical treatments, which aim at eliminating diseases at their cause or alleviating diseases by restoring homeostasis (Fig. 13.3).

13.2.3.2 Elements of Homeostatic Medicine

Homeostatic medicine comprises three key steps: (1) to understand the mechanism of homeostasis regulation and its role in maintaining health; (2) to analyze the causes and intervention factors of homeostasis imbalance in the process of disease; and (3) to consolidate the information gained in the first two steps and restore the homeostasis of the body through reasonable interventions, thereby promoting the maintenance of health and the treatment of disease. Given the critical role of

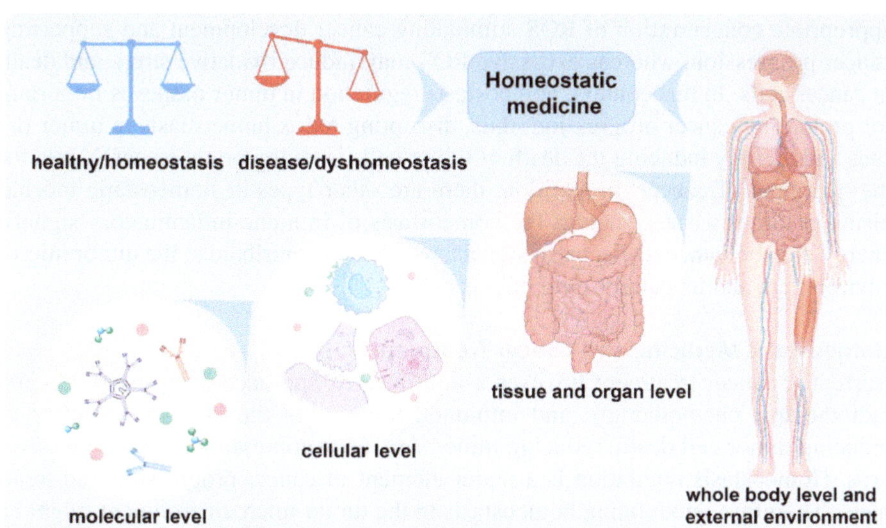

Fig. 13.3 Definition and elements of homeostatic medicine

homeostasis in health and disease, homeostatic medicine has many potential applications in a wide variety of pathological conditions.

13.2.4 Role of Homeostatic Medicine in Body Systems

13.2.4.1 Homeostatic Medicine and Cancer Progression and Treatment

Homeostatic Imbalance and Cancer Progression

Cancer progression is closely associated with the dyshomeostasis. Proto-oncogenes are a group of genes that play important roles in cell proliferation and differentiation. Typically, proto-oncogenes are strictly regulated by another group of genes, known as tumor suppressor genes, and the dynamic balance between them is helpful to maintain the number of cells and normal physiological functions *in vivo*. However, gene mutations, under the influence of specific stimulating factors, can disrupt this balance, and the overexpression of proto-oncogenes may lead to abnormal cell proliferation and tumor formation. Furthermore, apoptosis is a form of programmed cell death that plays a key role in embryonic development and the dynamic homeostasis of tissues and cells. Accordingly, the homeostasis regulation of apoptosis has opposing roles in tumor development. On the one hand, apoptosis inhibits tumor progression through the recognition of tumor cells and the subsequent execution of programmed cell death. On the other hand, apoptosis can also promote tumor progression by stimulating repair and regenerative responses in the tumor microenvironment. Thus, promoting or inhibiting apoptosis alone is not an effective strategy for cancer treatment [6]. Moreover, redox homeostasis is involved in tumor progression. Reactive oxygen species (ROS) play an opposing role in cancer, with an appropriate concentration of ROS stimulating cancer development and supporting cancer progression, whereas excessive ROS may induce oxidative stress and death in cancer cells. In this context, homeostasis regulation in tumor tissues is important for promoting cancer progression; thus, disrupting redox homeostasis in tumor tissues and thereby inducing the death of tumor cells is an important research area for the treatment of cancer. In addition, there are other types of homeostatic mechanisms in tumor tissues, such as the homeostasis of immune-inflammatory signals, energy metabolism, and electrical signals, which also contribute to the tumor microenvironment and its pathological effects [7].

Homeostatic Medicine and Cancer Treatment

Currently, cancer treatment involves a multifaceted approach comprising surgery, radiotherapy, chemotherapy, and immunotherapy, with the primary objective of inducing tumor cell death, reducing tumor size, and suppressing tumor aggressiveness. Homeostasis regulation is a major element of cancer progression and treatment. Therefore, modulating homeostasis in the tumor microenvironment might be a viable therapeutic strategy. Homeostatic medicine integrates the concepts of homeostasis regulation with cancer treatment by either restoring the homeostasis in normal tissues to increase their resistance to tumors or disrupting the homeostasis in tumor tissues to inhibit their growth. This approach has the potential to improve therapeutic effect while minimizing side effects, thereby obtaining twice the efficacy with half the effort. The use of pro-oxidants and antioxidants in cancer is a good example of this approach. In the early stage of cancer, redox homeostasis is disrupted, and oxidative stress promotes cancer progression. At this point, the use of antioxidants tends to inhibit cancer [8]. However, at the advanced stage of cancer, the antioxidant stress capacity of tumor tissues is augmented in response to oxidative stress-induced tumor cell death, so that pro-oxidants can be used to disrupt the redox homeostasis of tumor tissues and kill tumor cells [9]. Therefore, the targeted regulation of homeostasis, depending on the tumor stage, might be an effective option. Inorganic nitrates can increase the sensitivity of oral squamous cell carcinoma to chemotherapy by decreasing the expression of REDD1, and this modulatory effect may be linked to improved NO homeostasis in hypoxic tumor tissues [10]. Inorganic nitrates have also been found to exert relatively good protective effects against the side effects of radiotherapy for cancer treatment, which can effectively alleviate radiation-induced salivary gland dysfunction and colon damage in rats subjected to whole-body radiation by modulating the homeostasis of the gut microbiome [11]. This evidence indicates that inorganic nitrates have the potential to enhance the postoperative life quality in cancer patients undergoing radiotherapy. Furthermore, at lower radiation doses, such as in CBCT, the inhibitory effect of inorganic nitrates on oxidative stress induced by radiation injury has been demonstrated [12]. Whether the removal of all cancer tissues from the body should be the sole therapeutic goal remains a matter of debate. Even when there are tumors in the body, controlling tumor growth through treatment to maintain homeostasis should be a major therapeutic goal. In contrast, aggressively pursuing the elimination of all tumors through treatment can negatively affect the normal function of the body, thus significantly reducing the quality of life and potentially increasing mortality.

13.2.4.2 Homeostatic Medicine and Cardiovascular Disease

Homeostatic Imbalance and the Progression of Cardiovascular Disease

Homeostasis regulation is integral to the health and disease of the cardiovascular system. Vascular wall cells exhibit high sensitivity to changes in the mechanical environment. In healthy blood vessels, dynamic homeostasis is maintained at multiple levels via negative feedback regulation to maintain biomechanical stability. Conversely, the progression of cardiovascular disease is usually associated with the overregulation of biomechanical homeostasis or the destabilization of positive feedback [13]. The cardiovascular system adapts to different changes in the internal and external environments of the body through functional and structural changes. For example, a sustained increase in cardiac output leads to dilation of the central artery, thereby decreasing vascular resistance to blood flow and subsequently reducing cardiac workload. Alternatively, angiogenic signals are activated at sites of active tissue growth or enhanced local metabolic activity, thereby increasing the local blood supply to meet demand. This regulatory effect contributes to the maintenance of homeostasis. However, long-term vasoconstriction or vasodilation induces alterations in the structure of blood vessels, excessive dilation, and thinning of the blood vessels, which can lead to aneurysm or arterial rupture. To enhance vascular stress intensity, blood vessels begin to develop fibrosis. Eventually, excessive fibrosis of the blood vessels leads to vascular sclerosis and hemadostenosis. These changes cause atherosclerosis and hypertension, which further increase stress on the cardiovascular system. This abnormal positive feedback and tissue remodeling constitute one of the pathogenic mechanisms underlying cardiovascular disease [14].

Homeostatic Medicine and the Treatment of Cardiovascular Disease
Currently, the treatment of cardiovascular disease mainly focuses on alleviating the symptoms of patients, including strategies such as reducing the blood volume by using diuretics, limiting vasoconstriction by inhibiting the renin-angiotensin pathway and calcium channels, and lowering cholesterol synthesis to alleviate atherosclerosis by using statins. Although these treatments are common in clinical practice and have demonstrated a certain efficacy, many of these medications may cause side effects that limit their effects. Organic nitrates are commonly used as antihypertensive drugs, which act as NO donors to regulate vascular endothelial functions and serve as vascular smooth muscle relaxants to achieve antihypertensive effects. However, long-term use of organic nitrates leads to decreased efficacy and can also result in endothelial dysfunction and increased long-term cardiovascular risk [15].

Ideally, homeostatic medicine should comprehensively consider the mechanisms of vascular biomechanics, cell signal transduction, immunobiology, and other aspects of homeostasis regulation to avoid the disruption of homeostasis. By means of identifying and blocking therapeutic targets of disease processes, homeostatic medicine improves the state of a disease by restoring the dynamic balance of the cardiovascular system. Inorganic nitrates can maintain NO homeostasis through the nitrate-nitrite-NO pathway and have shown beneficial effects in a variety of cardiovascular diseases. Notably, a significant antihypertensive effect was still observed after one year of use of inorganic nitrate, indicating that inorganic nitrates do not result in the same drug resistance as organic nitrates. Moreover, the use of inorganic nitrates did not cause adverse effects, such as syncope or postural hypotension. Through the entero-salivary circulation, inorganic nitrates can be maintained at a stable concentration *in vivo* for long periods, ensuring a continuous supply of NO through the nitrate-nitrite-NO pathway. The advantageous antihypertensive effects of inorganic nitrates may be attributed to these mechanisms. In addition, compared to healthy volunteers, more significant antihypertensive effects of inorganic nitrates were observed in patients with hypertension and endothelial dysfunction, suggesting that inorganic nitrates may contribute to the amelioration of disturbances in endothelial homeostasis [16].

13.2.4.3 Homeostatic Medicine and Metabolic Diseases

Homeostatic Imbalance and the Progression of Metabolic Diseases

Energy homeostasis is mainly regulated by the central nervous system and notably involves the neuroendocrine center located in the hypothalamus and the solitary nucleus at the lower brain stem. These neuroendocrine centers exert physiological control over metabolism through the secretion of thyrotropin-releasing hormone and corticotropin-releasing hormone, but these centers are also modulated by 5-hydroxytryptamine (5-HT), thereby regulating food craving. A deficiency of 5-HT is associated with polyphagia and obesity, whereas elevated central 5-HT levels are linked to anorexia and reduced energy intake [17]. In addition, interactions between neurotransmitters of the central and peripheral nervous systems, including leptin, ghrelin, insulin, and cholecystokinin, are important elements of energy homeostasis. Leptin is secreted primarily by adipocytes, and its secretion is positively correlated with the amount of body fat. Insulin is secreted by pancreatic β-cells in response to elevated blood glucose, which promotes glucose consumption and storage, thereby lowering blood glucose. Increased secretion of these hormones typically indicates an excess in the energy reserves of the body and thus can also reduce appetite by regulating the central neuroendocrine system. In contrast, ghrelin is produced primarily in the stomach, and its secretion is increased during hunger, which enhances appetite [18].

Obesity is a metabolic disorder characterized by the excessive accumulation of body fat and is associated with an increased risk of various diseases. It is influenced by multiple regulatory mechanisms, including genetics, viral infection, insulin resistance, inflammation, the gut microbiome, circadian rhythms, and hormonal regulation. Metabolic homeostatic disorders are the primary causes of obesity. Normally, the feeding behavior of the human body is dynamically balanced. In other words, appetite increases when energy reserves decrease, resulting in the food intake needed to meet physiological demands. However, obese patients frequently display hedonic hunger, meaning that they may consume palatable foods beyond their energy requirements even when their energy reserves are sufficient. In addition, binge eating alters the sensitivity of the brain's dietary reward center to dopamine, further exacerbating the disruption of energy metabolism homeostasis and promoting the progression of obesity [19].

Insulin resistance is a pathological condition characterized by diminished sensitivity to insulin, leading to decreased efficacy of insulin-mediated glucose uptake and reduced glucose utilization by cells of the liver, skeletal muscle, adipose tissue, and other tissues. Under inflammatory conditions caused by infection, stress, or injury, tumor necrosis factors (TNF) reduce the sensitivity of fat and skeletal muscle to insulin, resulting in the allocation of more energy to the immune system. In addition, the expression of GLUT2 and glucokinase in pancreatic β-cells is suppressed under inflammatory conditions, resulting in decreased sensitivity to blood glucose levels and decreased insulin production. These regulatory effects impair homeostasis by reducing tissue sensitivity to insulin and systemic sensitivity to blood glucose. Long-term changes in this homeostasis may result in irreversible decline in insulin potency and decreased sensitivity to elevated blood glucose, ultimately leading to insulin resistance. Therefore, homeostasis regulation is a critical step in the reversal of metabolic diseases [20].

Homeostatic Medicine and the Treatment of Metabolic Diseases

The primary objective of homeostatic medicine for metabolic diseases is to understand the mechanisms underlying homeostatic imbalance in metabolic diseases, aiming to eliminate the imbalance between energy intake and metabolism and to restore altered equilibrium values. Inorganic nitrates have beneficial effects on metabolic diseases, which may be related to the promotion of the nitrate-nitrite-NO balance, thereby maintaining NO homeostasis and regulating microbial homeostasis. Inorganic nitrates can reduce body fat and improve glucose homeostasis in eNOS-deficient mice with impaired NO synthesis. Liver senescence and metabolic dysfunction are also major causes of metabolic diseases. In d-galactose-induced aging mice and naturally aging mice, daily nitrate intake can effectively restore the liver metabolic capacity by preventing senescence of liver cells and the degeneration of glucose and lipid metabolism in hepatic tissue [21]. Furthermore, by activating the nitrate-nitrite-NO pathway, inorganic nitrates can promote the conversion of white adipose tissue to brown adipose tissue. Inorganic nitrates can also reduce high-fat diet-induced obesity and ameliorate glucose and lipid metabolism disorders in mice by activating the NO pathway and regulating the gut microbiome [22].

Homeostatic Medicine and Immune and Infectious Diseases

Organisms are frequently confronted by challenges in the external environment. The immune system comprises the primary defense of the body against harmful external stimuli and antigens. Furthermore, it plays a crucial role in maintaining homeostasis through immune surveillance, sensing metabolic changes, and controlling the inflammation caused by external stimuli. The immune system can be divided into innate and adaptive immune systems. These two systems do not act independently but interact through a variety of cellular and humoral effectors, collectively contributing to the dynamic balance of the body's immunity. Disturbances of immune homeostasis are closely associated with disease, and disease progression often results from immune regulation failure [23].

Novel coronavirus pneumonia (COVID-19) is an infectious disease caused by a novel coronavirus (SARS-CoV-2) and characterized by acute respiratory failure. SARS-CoV-2 can enter cells after binding to angiotensin-converting enzyme 2 (ACE2). After infection, cells infected with SARS-CoV-2 are recognized by macrophages and dendritic cells. To clear virus-infected cells as quickly as possible, macrophages produce a large number of proinflammatory cytokines and chemokines, leading to the further recruitment of inflammatory cells, such as neutrophils, to the site of infection. This excessive release of proinflammatory cytokines can result in cytokine storms that can lead to tissue damage, organ failure, and death [24]. Severe symptoms of COVID-19 are most common in elderly patients and are often accompanied by complications, whereas younger people tend to have milder symptoms. This is because the immune system undergoes a complex maturation process from birth to adulthood, with immune defenses being more mature in young adults and homeostasis regulation changing with the aging of the body. Compared with younger patients, elderly patients exhibit a diminished capacity to respond to and eliminate viral infections, while their inflammatory responses to infections tend to be more active. This characteristic of the immune system is the main reason for the high severity and mortality of COVID-19 in elderly patients [25].

In addition, an inadequate immune response, exemplified by HIV infection which specifically targets $CD4^+$ T cells, results in the depletion of the body's immune capabilities, ultimately leading to various infectious and neoplastic diseases that can cause multiple organ failure and death. Conversely, an excessive or abnormal immune response can cause allergic reactions or autoimmune disease. Thus, restoring immune system homeostasis is essential for the treatment of immunological disorders. B-cell hyperactivation is an important pathogenic mechanism that has been implicated in various autoimmune diseases, such as systemic lupus erythematosus, Sjögren's syndrome, and rheumatoid arthritis, and it is associated with disturbances in the signals that regulate B cells. The inhibition of B-cell activating factor (BAFF), hypoxia-inducible factor (HIF)-1α, and TNF receptor-associated factor (TNFR)-3 contributes to the restoration of B-cell metabolic activity, thereby reducing the viability of self-reactive B cells and attenuating the symptoms of autoimmune diseases [26].

Pretreatment with dietary nitrates attenuates the inflammatory response of leukocytes to acute chemokine-induced inflammation by delivering NO to the vascular system through the nitrate-nitrite-NO pathway [27]. In mice susceptible to atherosclerosis, inorganic nitrate intake rescued NO homeostasis and exerted significant anti-inflammatory effects by inhibiting neutrophil activation and increasing IL-10 levels to inhibit acute and chronic inflammation. Nitrates reduced the number of macrophages in atherosclerotic plaques and inhibited their inflammatory activity by reversing NO deficiency caused by endothelial dysfunction [28]. In the context of liver oxidative stress injury induced by ischemia-reperfusion, inorganic nitrate intake increased NO levels in the plasma and liver and relieved liver oxidative stress by upregulating nuclear factor erythroid 2-related factor 2 (NRF2)-related molecules and increasing antioxidant enzyme activity [29]. Furthermore, microbial homeostasis in the human body plays an equally important regulatory role in infectious diseases and inflammatory processes. Nitrates serve as an NO donor and have been demonstrated to exert inhibitory effects on pathogenic bacteria. Inorganic nitrates have been shown to mitigate dextran sodium sulfate-induced enteritis by regulating homeostasis of the gut microbiome, increasing colon length, and maintaining the body weight of mice [30].

13.2.5 Future Directions and Prospects of Homeostatic Medicine

Homeostasis regulation plays an important role in health maintenance and disease prevention in different systems. Normal homeostasis regulation is fundamental for maintaining health, and disruption of this balance can cause body dysfunction and ultimately lead to disease. Restoring homeostasis is therefore a key strategy used to reverse disease states. Homeostatic medicine integrates the mechanisms of homeostasis regulation at the molecular, cellular, organ, and system levels. By understanding the regulatory mechanism of homeostasis in health and disease states, the imbalance in homeostasis can be reversed through reasonable intervention. In this way, homeostatic medicine contributes to the maintenance of health and the treatment of disease. Homeostatic medicine is a novel medical system developed from previous medical concepts with potentially wide-ranging prospects, and it is expected to provide many new topics and strategies for biomedical research and therapeutic interventions.

Compliance with Ethical Standards All applicable international, national, and/or institutional guidelines for the care and use of animals were followed.

References

1. Gross CG. Three before their time: neuroscientists whose ideas were ignored by their contemporaries. Exp Brain Res. 2009;192:321–34.
2. Cooper SJ. From Claude Bernard to Walter Cannon. Emergence of the concept of homeostasis, Appetite. 2008;51:419–27.
3. Maiese K. Cellular balance, genes, and the Huang Ti Nei Ching Su Wen. Curr Neurovasc Res. 2006;3:247–8.
4. Goldstein DS, Kopin IJ. Homeostatic systems, biocybernetics, and autonomic neuroscience. Auton Neurosci. 2017;208:15–28.
5. Goodman L. Regulation and control in physiological systems: 1960-1980. Ann Biomed Eng. 1980;8:281–90.
6. Morana O, Wood W, Gregory CD. The Apoptosis Paradox in Cancer. Int J Mol Sci. 2022:23.
7. Sheth M, Esfandiari L. Bioelectric Dysregulation in Cancer Initiation. Promotion, and Progression, Front Oncol. 2022;12:846917.
8. Panieri E, Santoro MM. ROS homeostasis and metabolism: a dangerous liason in cancer cells. Cell Death Dis. 2016;7:e2253.
9. Conklin KA. Chemotherapy-associated oxidative stress: impact on chemotherapeutic effectiveness. Integr Cancer Ther. 2004;3:294–300.
10. Feng Y, Cao X, Zhao B, et al. Nitrate increases cisplatin chemosensitivity of oral squamous cell carcinoma via REDD1/AKT signaling pathway. Sci China Life Sci. 2021;64:1814–28.
11. Wnag W, Hu L, Chang S, et al. Total body irradiation-induced colon damage is prevented by nitrate-mediated suppression of oxidative stress and homeostasis of the gut microbiome. Nitric Oxide. 2020;102:1–11.
12. Chang S, Hu L, Xu Y, et al. Inorganic Nitrate Alleviates Total Body Irradiation-Induced Systemic Damage by Decreasing Reactive Oxygen Species Levels. Int J Radiat Oncol Biol Phys. 2019;103:945–57.
13. Humphrey JD, Schwartz MA. Vascular Mechanobiology: Homeostasis, Adaptation, and Disease. Annu Rev Biomed Eng. 2021;23:1–27.
14. Takebe T, Imai R, Ono S. The Current Status of Drug Discovery and Development as Originated in United States Academia: The Influence of Industrial and Academic Collaboration on Drug Discovery and Development. Clin Transl Sci. 2018;11:597–606.
15. Tarkin JM, Kaski JC. Vasodilator Therapy: Nitrates and Nicorandil. Cardiovasc Drugs Ther. 2016;30:367–78.

16. Kapil V, Khambata RS, Robertson A, et al. Dietary nitrate provides sustained blood pressure lowering in hypertensive patients: a randomized, phase 2, double-blind, placebo-controlled study. Hypertension. 2015;65:320–7.
17. Koliaki C, Liatis S, Dalamaga M, et al. The Implication of Gut Hormones in the Regulation of Energy Homeostasis and Their Role in the Pathophysiology of Obesity. Curr Obes Rep. 2020;9:255–71.
18. Makris MC, Alexandrou A, Papatsoustsos EG, et al. Ghrelin and Obesity: Identifying Gaps and Dispelling Myths. A Reappraisal, In Vivo. 2017;31:1047–50.
19. Kessler RM, Hutson PH, Herman BK, et al. The neurobiological basis of binge-eating disorder. Neurosci Biobehav Rev. 2016;63:223–38.
20. Kotas ME, Medzhitov R. Homeostasis, inflammation, and disease susceptibility. Cell. 2015;160:816–27.
21. Wang H, Hu L, Li L, et al. Inorganic nitrate alleviates the senescence-related decline in liver function. Sci China Life Sci. 2018;61:24–34.
22. Ma L, Hu L, Jin L, et al. Rebalancing glucolipid metabolism and gut microbiome dysbiosis by nitrate-dependent alleviation of high-fat diet-induced obesity. BMJ Open Diabetes Res Care. 2020:8.
23. Kenedy MA. A brief review of the basics of immunology: the innate and adaptive response. Vet Clin North Am Small Anim Pract. 2010;40:369–79.
24. Zhou Y, Fu B, Zheng X, et al. Pathogenic T-cells and inflammatory monocytes incite inflammatory storms in severe COVID-19 patients. Natl Sci Rev. 2020;7:998–1002.
25. Fulop T, Larbi A, Dupuis G, et al. Immunosenescence and Inflamm-Aging As Two Sides of the Same Coin: Friends or Foes? Front Immunol. 1960;2017:8.
26. Mubariki R, Vadasz Z. The role of B cell metabolism in autoimmune diseases. Autoimmun Rev. 2022;21:103116.
27. Jadert C, Retersson J, Massena S, et al. Decreased leukocyte recruitment by inorganic nitrate and nitrite in microvascular inflammation and NSAID-induced intestinal injury. Free Radic Biol Med. 2012;52:683–92.
28. Khambata RS, Ghosh SM, Rathod KS, et al. Antiinflammatory actions of inorganic nitrate stabilize the atherosclerotic plaque. Proc Natl Acad Sci U S A. 2017;114:E550–9.
29. Li S, Jin H, Sun G, et al. Dietary Inorganic Nitrate Protects Hepatic Ischemia-Reperfusion Injury Through NRF2-Mediated Antioxidative Stress. Front Pharmacol. 2021;12:634115.
30. Hu L, Jin L, Xia D, et al. Nitrate ameliorates dextran sodium sulfate-induced colitis by regulating the homeostasis of the intestinal microbiota. Free Radic Biol Med. 2020;152:609–21.

MIX
Papier aus verantwortungsvollen Quellen
Paper from responsible sources
FSC® C105338
FSC
www.fsc.org

If you have any concerns about our products,
you can contact us on
ProductSafety@springernature.com

In case Publisher is established outside the EU,
the EU authorized representative is:
Springer Nature Customer Service Center GmbH
Europaplatz 3, 69115 Heidelberg, Germany

Printed by Libri Plureos GmbH
in Hamburg, Germany